Using R for
Bayesian Spatial and
Spatio-Temporal
Health Modeling

Using R for Bayesian Spatial and Spatio-Temporal Health Modeling

Andrew B. Lawson

CRC Press
Taylor & Francis Group
Boca Raton London New York

CRC Press is an imprint of the
Taylor & Francis Group, an **informa** business
A CHAPMAN & HALL BOOK

First edition published 2021
by CRC Press
6000 Broken Sound Parkway NW, Suite 300, Boca Raton, FL 33487-2742

and by CRC Press
2 Park Square, Milton Park, Abingdon, Oxon, OX14 4RN

© 2021 Taylor & Francis Group, LLC

CRC Press is an imprint of Taylor & Francis Group, LLC

ISBN: 9780367490126 (hbk)
ISBN: 9781003043997(ebk)

Typeset in CMR10
by KnowledgeWorks Global Ltd.

Contents

Preface

Bayesian methods for disease mapping have reached a maturity with a wide variety of software and associated methods. With the development of a wider range of posterior sampling algorithms and various posterior approximation methods, some sophistication in analysis can be within reach of a wide range of potential users.

In this work the aim is to provide an entry level account of the use of R to analyze spatial health data via Bayesian Hierarchical modeling (BHMs). A range of software is now available to achieve this task. Here, the focus is on a small range of flexible R packages that can provide the tools for this: BRugs/OpenBUGS, CARBayes, Nimble, and INLA. The focus of the work is on the gradual introduction of different models with different complexity levels and demonstration of how to fit these models with the relevant software. The comparison of software forms an extension to the discussion in Chapter 15 of Lawson (2018), and further provides direct examples of code, their output, and discussion of the relative advantages and disadvantages.

There are many people who have helped in the production of this work. In particular, Diba Khan (CDC), Matt Bozigar, Joanne Kim and Daniel Baer (@MUSC), and Chris Paciorek for Nimble support. I must also thank those at CRC press for great help in finalizing the work. In particular, Rob Calver and Lara Spieker for general production support and Vishali Singh for Latex help.

Finally I would like to again acknowledge the enduring support of my family, and, in particular, Pat for her understanding during the sometimes difficult activity of book writing, especially during a pandemic.

Andrew B. Lawson
Charleston, South Carolina,
USA
2020

Biography

Professor Andrew B. Lawson

Dr Lawson is Professor of Biostatistics in the Division of Biostatistics and Bioinformatics, Department of Public Health Sciences, College of Medicine, MUSC and is an MUSC Distinguished Professor Emeritus and ASA Fellow. His PhD was in Spatial Statistics from the University of St. Andrews, Scotland, United Kingdom.

He has over 190 journal papers on the subject of spatial epidemiology, spatial statistics, and related areas. In addition to a number of book chapters, he is the author of 10 books in areas related to spatial epidemiology and health surveillance. The most recent of these is Lawson, A.B. et al (eds) (2016) *Handbook of Spatial Epidemiology*. CRC Press, New York, and in 2018 a 3rd edition of *Bayesian Disease Mapping; hierarchical modeling in spatial epidemiology* CRC Press. He has acted as an advisor in disease mapping and risk assessment for the World Health Organization (WHO) and is the founding editor of the Elsevier journal Spatial and Spatio-temporal Epidemiology. Dr Lawson has delivered many short courses in different locations over the last 20 years on Bayesian Disease Mapping with OpenBUGS, INLA, and Nimble, and more general spatial epidemiology topics.

Web site: http://people.musc.edu/~abl6/

List of Tables

1

Introduction and Datasets

Bayesian spatial health modeling, sometimes also known as Bayesian disease mapping, has matured to the extent that a range of computational tools exist to enhance end user's ability to analyze and interpret the variations in disease risk commonly found in human and animal populations. This has been enhanced by the easy availability of geographical information systems (GIS) such as ArcGIS or Quantum GIS (QGIS). While a variety of software platforms host specialist programs, there is now a large body of software available for the R programming environment, and given the general accessibility of this free software platform it is advantageous to consider the integration of analyses on this platform. A range of Bayesian modeling software is now available on R which can be used for spatial health modeling. Within R, it is possible to process geo-referenced data, including GIS-based information such as adjacency of regions and polygon neighborhoods. With the standard capabilities of R for descriptive statistics and R's extensive plotting facilities, it is easy to explore geo-referenced data.

With the addition of a range of packages that can fit Bayesian Hierarchical models (BHMs) it is now possible to carry out both exploration and analysis of spatial health data in one environment. In addition, a selection of packages that fit BHMs can also fit spatial dependence structures and so address different kinds of spatial confounding of risk. This means that state-of-the-art models can be fitted and analyzed.

Estimation in Bayesian HMs is based on a parameter posterior distribution, which is a function of a data likelihood and prior distributions for model parameters. Bayesian HMs are often relatively sophisticated and require approximation methods to address parameter estimation. Two major approaches to this approximation are commonly found: (1) posterior sampling where an iterative algorithm is used to provide a sample of parameter values that can be summarized to yield estimates, (2) numerical approximation to an integral, based on an approximate form of the posterior distribution. Posterior sampling is often carried out using Monte Carlo methods and a set of methods called Markov chain Monte Carlo (McMC) have been developed to facilitate this estimation approach. Essentially in this approach the posterior distribution is approximated by samples, and sampled values are used to summarize the posterior quantities of interest (such as mean, variance etc). Numerical approximation of the posterior distribution commonly involves a Laplace approximation which matches a Gaussian distribution to the form of the posterior distribution and provides estimates based on this approximation. Numerical approximation can be computationally advantageous especially in large-scale problems (big data) where sampling becomes inefficient. On the other hand McMC can provide a large amount of information concerning the form of the posterior distribution, which is not usually available immediately from numerical approximation.

For the application of BHMs to disease mapping, there is an extensive literature now available (e.g., Breslow and Clayton, 1993; Besag et al., 1991; Leroux et al., 2000; Blangiardo et al., 2013 and Lawson, 2018 for a review). Often, this area is termed Bayesian Disease Mapping (BDM) and this acronym will be used extensively here. A major issue that arises when considering the model-based approach to disease mapping is the assumption of the spatial continuity of risk. Disease risk, as displayed as incidence in small areas, is dependent on the underlying population at risk of the disease. As this population is discrete in nature, in that subjects must exist for disease to occur, then disease risk will also be discrete. At the individual level a subject will have a binary outcome (disease/no disease) and when aggregated to a population then a count of disease arises. At the finest spatial scale an individual subject could have an address location associated with them. This could be a residential address for a person or a farm for an animal (say). Essentially the location is a unique identifier. At this scale the location is stochastic and a point process of cases and controls will arise. Once aggregated to arbitrary small areas (census tracts, post codes, provinces, etc.) then counts within areas of cases, and of controls, will arise. Often, if the population is large relative to the probability of disease, i.e., in the case of a relatively rare disease, then a Poisson count data model for the cases is often assumed. When a smaller finite population occurs then a binomial model is more appropriate with case count modeled in relation to the case plus control total population. These data models which assume independent contributions from each subject can be justified based on conditional independence within the BHM hierarchy.

Although populations are discrete, it is also the case that a choice of risk model can be made whereby components of risk are assumed to be continuous. For example, it is reasonable to assume that environmental stressors such as pollution levels are spatially and temporally continuous. Often BDM models consist of fixed (predictor) effects and random effects. These represent observed outcome confounders and unobserved confounding respectively. The choice of random effect to model unobserved confounding can lead to markedly different model paradigms. If the unobserved correlated confounding is continuous then it would be justified to consider continuous spatial/temporal random effects (such as spatial/temporal Gaussian processes). However, instead, it is more usual to assume Markov random field (MRF) models, whereby only local neighborhood dependence defines the correlated effect. Given the discrete nature of small areas, this would usually seem to be a natural approach.

In this work both spatial BDM models and spatio-temporal examples will be considered. Spatio-temporal (ST) data arises naturally in disease mapping studies, as all cases of disease have associated a date and/or time of diagnosis. Aggregation of cases into time periods and small areas leads to spatiotemporal count sequences. Spatial BDM models can be extended relatively easily to deal with ST data in the form of counts in time periods and small areas. Special models can be developed for ST data variants, such as spatial survival modeling, spatial recurrent event, or longitudinal modeling.

To set the scene with respect to historical development, a brief, and by no means exhaustive, overview of the development of approaches to BDM is presented here. Early examples of Bayesian models were (Clayton and Kaldor, 1987) using empirical Bayes approximation, the use of MRF models for (transformed) counts include Cressie and Chan (1989), Bayesian Poisson data models with conditional autoregressive random effects and MCMC (Besag et al., 1991), space-time models (Bernardinelli et al., 1995; Waller et al., 1997; Knorr-Held, 2000). Some more recent developments with special focus are: (a) multivariate disease mapping (Gelfand and Vounatsou, 2003, Jin et al., 2005; Jin et al., 2007; Martinez-Beneito et al., 2017), (b) spatial survival modeling (Osnes and Aalen, 1999; Henderson et al., 2002; Banerjee et al., 2003; Cooner et al., 2006; Zhou et al., 2008; Lawson and Song, 2009; Banerjee, 2016; Onicescu et al., 2017b; Onicescu et al., 2017a), (c) spatial longitudinal modeling (Lawson et al., 2014), and (d) infectious disease modeling (Sattenspiel and Lloyd, 2009; Lawson and Song, 2010; Chan et al., 2010; Lawson et al., 2011; Corberán-Vallet, 2012).

Bayesian hierarchical modeling (BHM) often requires sophisticated numerical methods to be employed to provide estimates of relevant parameters. To this end, software packages have been developed on the R platform that either implement posterior sampling via MCMC, or numerical integration using Laplace approximation. While a number of variants exist of these main computational approaches, in this work the focus is on the main packages which implement posterior sampling and Laplace approximation for BDM. These consist of WinBUGS/OpenBUGS, nimble, CARBayes, which use posterior sampling. STAN and jags are not implemented here as they have been used less frequently. Jags does not have spatial prior distributions and STAN only recently implemented these models (see e.g., package brms and the cor_car function, Morris et al., 2019). For numerical approximation, the Laplace approximation as implemented in the R-INLA package is reviewed.

This work is closely tied to the preceding volume on Bayesian Disease Mapping by Lawson (2018). The intention with this work is to provide direct exposure to the software that can be used to implement BHMs in the context of disease mapping. To this end, detailed discussion of models and their justification is not attempted, but reference is made to book sections in Lawson (2018) where the models and computational approaches are described in more detail. In addition, as a cloud resource a variety of examples of model code in different forms (BUGS, INLA, nimble, CARBayes) are available on GitHUB: https://github.com/Andrew9Lawson?? . Two different branches can be found at this site. A nimble code branch and other software branch (OpenBUGS,

INLA, CARBayes). Examples used in this work can also be found at this site. Additional resources can also be found at http://people.musc.edu/~abl6/. The software version of R used in this work is 3.6.2.

1.1 Datasets

Here, I describe the basic datasets that will be analyzed in this work. As a comparison of software is to be made I have limited the range of datasets to a small number so that direct comparison is facilitated.

Spatial data

(1) All ICD code respiratory cancers at county level in South Carolina, USA, for the year 1998. This represents a broad spectrum of etiology and while risk is relatively low the counts are not sparse (i.e., no zero counts in counties). The data consist of counts of cases and expected rates computed from the statewide population rate. I assume indirect standardization. The standardized incidence ratio (SIR) is computed from the ratio : count/expected rate. Figure 1.1 displays the marginal density estimate of the standardized incidence ratio. Figure 1.2 displays the mapped SIR for these data.

(2) Multivariate example: asthma, COPD, and angina incidence in counties of Georgia, USA for the year 2005. These data were originally analyzed in Lawson (2018), Chapter 10. These data are considered to possibly be correlated in that their spatial incidence distribution may show similar clustering or other artifacts of disease incidence. Figure 1.3 displays the SIRs (relative risk estimates) for the incidence of these respiratory and coronary outcomes in the counties of Georgia USA for year 2005. The expected rates were calculated from statewide incidence rate for each disease separately.

(3) Low birth weight and very low birth weight events in 2007 in the counties of Georgia USA. This example includes an aggregate count outcome with a binomial model with 5 predictors and is featured in the variable selection section of Chapter 16. This is used as an example of aggregate-level regression where selection of variables is a focus.

(4) Cross-sectional time period (single biweekly) of foot and Mouth disease (FMD) within parishes during the epidemic in North West England in 2001. Counts of infected premises are available as well as total numbers of premises. Figure

Spatio-temporal data

Outcomes in the form of counts within small areas associated with a sequence of discrete time periods are also often a focus of analysis. In this situation the focus is usually on both spatial, temporal, as well as spatio-temporal (ST) interaction effects.

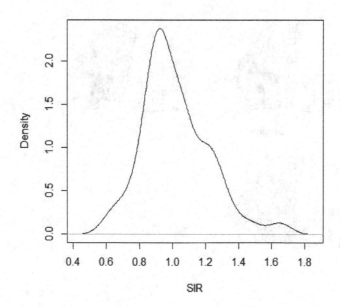

FIGURE 1.1
Standardized Incidence ratio for all respiratory cancers in South Carolina
counties in 1998: marginal density estimate using a default bandwidth.

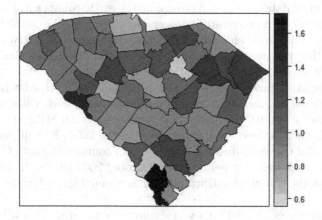

FIGURE 1.2
Standarized incidence ratio for all respiratory cancers in 1998, South Carolina
counties, USA.

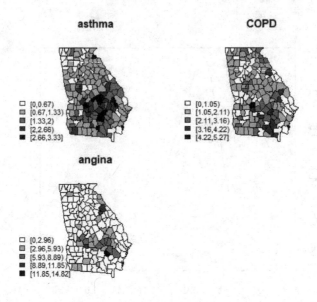

FIGURE 1.3
Relative risks (SIRs) for asthma, COPD, and angina incidence in the state of
Georgia, USA, for the year 2005.

(1) The main dataset that will be explored is yearly counts within counties of
South Carolina for all respiratory cancers for the years 2011–2016. Figure 1.4
displays the SIRs for each year in sequence. The expected rate was obtained
by computing the overall state wide rate for the disease for that year and
applying it to the county population in that year. It is clear that a few
areas of the state have high risk (>1.5 SIR) and the north eastern group of
counties shows repeated high risk across most years, even when the overall
rate decreases. In 2016 the highest risk area exceeds 2.0 SIR.

(2) SC influenza C+ve notifications during the 2004–2005 flu season. The
data consist of C+ve notification counts within counties of South Carolina for
a sequence of 13 biweekly periods from October 2004 to March 2005. Figure
1.5 displays the counts of notifications for four counties within the state.

(3) County-level Covid-19 daily case counts for the state of South Carolina:
January 22nd–April 12th and April 2nd to June 29th. The data was obtained
from the New York Times GitHub repository: https://github.com/nytimes/
covid-19-data. These data consist of confirmed daily case counts, supplied by

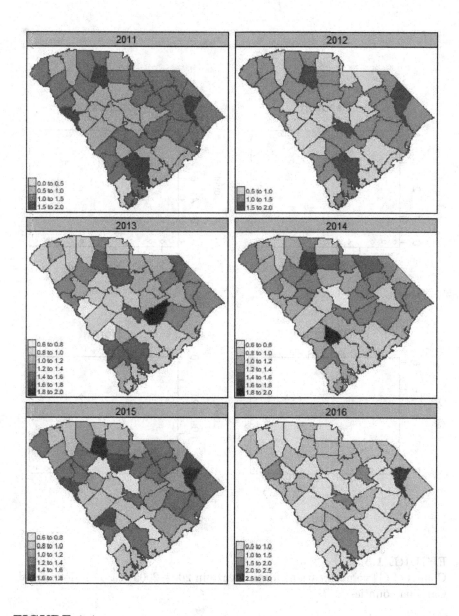

FIGURE 1.4

All respiratory cancers in South Carolina counties: standardized incidence ratios (SIRs) for the years 2011 to 2016.

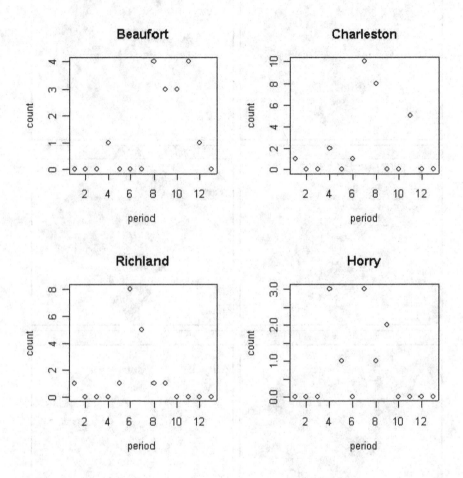

FIGURE 1.5
Counts of C+ve notifications for influenza in 2004–2005 season in four South
Carolina counties.

the relevant health department, and deaths as confirmed from the National Center for Health Statistics (NCHS). Figure 1.6 displays the case and death counts for a selection of four counties for the earlier period: January 22nd–April 2nd. Due to reporting biases relating to weekends, a smoothed 3-day average is also examined in Chapter 18.

Spatial survival data

Outcomes pertaining to the course of disease progression are often characterized by a time to event outcome (rather than incidence). For example, date of diagnosis is a common time end point, whereas date of vital outcome is a further marker of progression. For cancer registry data often both end points are known. Survival data of this kind are often complicated by the fact that some individuals may not reach the end point during the observation period. For example, in cancer registry studies, individuals usually have a date of diagnosis, when they enter the registry, but may not have a vital outcome (death) during the period of study. Those who don't have a vital outcome are considered to be censored, in that they didn't have the end point during the observation period. This is known as right censoring and is the commonest form of censoring in survival studies. Here, an example of cancer registry data is analyzed. The example used is of prostate cancer in the US state of Louisiana registry, which is one of the federally accredited SEER registries. Figure 1.7 displays the parishes of Louisiana which are the basis of the georeferencing of the survival data, i.e., each individual in the registry has a parish address and so the spatial information in this example is contextual: the parish is used as a contextual spatial effect in these data.

The data consists of all registered cases of prostate cancer diagnosed in Louisiana during the period 2007–2010. Vital outcome is also recorded within the time period of observation.

Case event data

Often individual-level patient-based data becomes available. Individual-level data is the fundamental data form that arises in spatial epidemiology and is the most fundamental level in most biomedical studies. The main difference in the case of disease mapping is that the individual subject has a geocoding defining their location (residential or otherwise). The finest level of geocoding is an exact address which offers a substitute for exposure or other locational effect. For example in a cluster study it may be important to assess the degree of exposure to a potential air pollution source. If so then exposure at a location near to the source could be important information associating case outcomes to air pollution insult. Equally, residential location could lead to evidence of a disease cluster. If cases are found to be located close to each other then a disease cluster might emerge. In both these examples residential address may be a surrogate for a common locationally specific effect. On the other hand, residential address may not be relevant when an exposure occurs during travel or in the workplace. Mesothelioma affected many shipyard workers in the eastern seaboard of the US in the early 20th century, but their workplaces (shipyards) were the exposure locale (Sanden and Jarvholm, 1991).

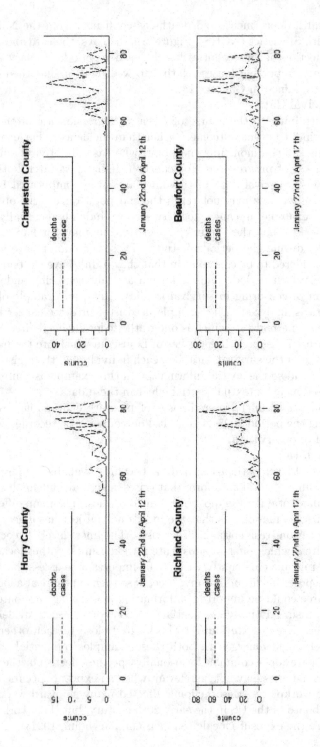

FIGURE 1.6
South Carolina Covid-19 case counts and deaths January 22nd to April 12th 2020: four counties shown.

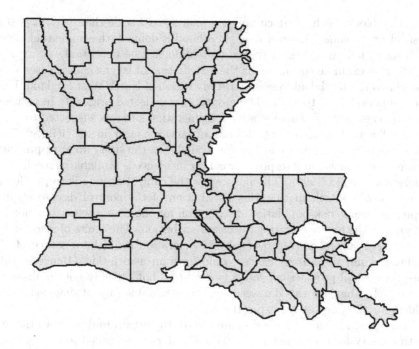

FIGURE 1.7
Louisiana parish map.

In another situation, geocoding could be at a more aggregate level. For example, it is now common that residential address may not be available due to patient confidentiality restrictions, and a geocoding unit above address level is available. In the US, below state level, aggregate geocoding could be at census, block/block group/tract, zip code, or county. In many countries, postal code municipality, and county or province levels are commonly available. In essence, an individual has an assigned aggregate geocode, and this becomes the locational tag for that individual. These geocodes are arbitrary geographic units, usually set up for administrative purposes, and usually unrelated to spatial variation in disease risk. Of course, health districts could be used to allocate health spending and so those districts could be related to disease incidence. Note that multiple geocodes at different spatial aggregation levels could be assigned to each individual and these are essentially contextual effects: the effect of lying within that geocode unit. The notion of "neighborhood" and its effect on individuals comes into play here (Kawachi and Berkman, 2003): if the geocode unit corresponds to a neighborhood then that unit could represent a neighborhood effect. However the "neighborhood" concept is difficult to precisely define, as many spatial scales could be assigned the term "neighborhood." A self-identified neighborhood could be a few streets, or a city block, both of which are at a finer spatial scale than (say) a postal district or zip code. However a neighborhood is defined, the contextual effect of lying within an aggregate spatial unit could still be important.

The first example considered is the distribution of larynx cancer cases with associated residential addresses in a study region of North West England, UK, for the period 1973 to 1984. These data were collected originally in relation to a "cluster alarm," whereby a local incineration site was suspected to have been influential in increasing the risk of larynx cancer in its vicinity. The incidence of a "control disease" is also available in the study area: respiratory cancer. Both larynx and respiratory cancer incidence is available as residential addresses (Diggle, 1990). The purpose of the control disease is to provide a *geographical* control, in that the distribution of the control disease should represent the at-risk population distribution for the case disease, but not be affected by the process creating the case distribution. The choice of respiratory cancer as a control is controversial as incineration could be a source of air pollutants and respiratory cancer could have an association. Hence in this case the control provides a *relative risk control,* but possibly not an absolute control. To augment this dataset a synthetic variable (age at diagnosis) was added, for the purposes of exposition only.

In the second example, we examine the birth certificate data from the SC birth registry for the period 1997–2009. The data is geocoded to county level and we have a range of measures on each birth that we can analyze in relation to selected outcomes. These measures include birthweight, race of mother, mother's age, county of residence, and a low birth weight (LBW) indicator. This is examined in Chapter 17. Figure 1.8   displays the county-level summary of these data for LBW in relation to total births. Higher crude ratios appear in relatively rural counties in SC. In what follows the individual-level LBW indicator will be analyzed with individual-level predictors predictors and contextual county-level effects.

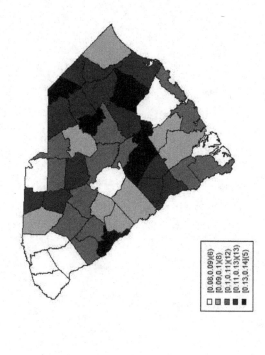

FIGURE 1.8
SC counties low birth weight(LBW) births 1997–2009: left to right: low birth weight counts, crude risk relative to total births

2

R Graphics and Spatial Health Data

In this chapter the basic approaches to visualizing spatial heath data on R will be addressed. As there are a range of R packages available to achieve visualization I will examine a subset of packages that are particulalry useful rather than attempting a review of all that exist. I will consider first three-dimensional surface visualization and surface drawing, followed by tools for thematic mapping and finally will consider some special health applications.

A measure at a spatial location is denoted z. Spatial data is usually available in the form of x,y,z coordinates, or at least the measure (z) will be associated with a spatial object (which may have an associated point x,y). A classic example of the former would be air pollution measures obtained at a network of measurement sites. An example of the latter would be a count of disease observed in a small area (such as zipcode or census tract). In the latter case the spatial object is a polygon.

2.1   Three-dimensional Surface Visualization

R has a wide range of packages for specialist plotting. Here I will focus on the akima and mba packages and the image, contour, persp, and filled.contour commands.

The akima package is a basic package which includes facilities for the interpolation of randomly located spatial data. The packages interpolate input data to a regular grid mesh which can then be used in standard plotting routines. akima uses a high order mathematical interpolator to achieve high accuracy. Several cubic spline interpolation methods for irregular and regular gridded data are available through this package, both for the bivariate case and univariate case. A bilinear interpolator for regular grids was also added for comparison with the bicubic interpolator on regular grids. The interp command can be used for irregularly spaced bivariate data.

```
library(akima)
asINT<-interp(x,y,z,linear=F,nx=200,ny=200)
image(asINT)
contour(asINT,add=T)
```

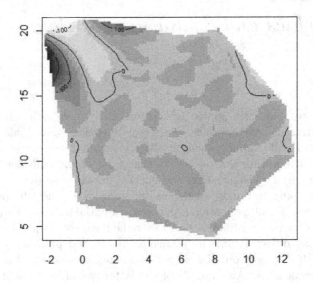

FIGURE 2.1
Bicubic spline plot using the akima package.

The above code specifies a bicubic spline interpolator with 200 x 200 grid
mesh. The resulting image and contouring is displayed in Figure 2.1.

Additional commands for perspective viewing and filled contouring are
persp(asINT) for wire frame viewing and filled.contour(asINT). Figure 2.2 displays the filled contour (level) plotting capabilities.

The package MBA has a more sophisticated interpolation approach: multilevel B-splines and can provide a more nuanced mapping experience. This
package includes functions to interpolate irregularly and regularly spaced data
using multilevel B-spline approximation (MBA). The functions call portions
of the SINTEF Multilevel B-spline Library which includes multiresolution
functionality, written by Øyvind Hjelle. Figure 2.3 displays the results of the
code below, which uses a 200 x 200 grid:

```
xyz<-data.frame(x,y,z)
mba.int <- mba.surf(xyz, 200, 200, extend=FALSE)$xyz.est
image(mba.int)
contour(mba.int,nlevels=5,add=T)
```

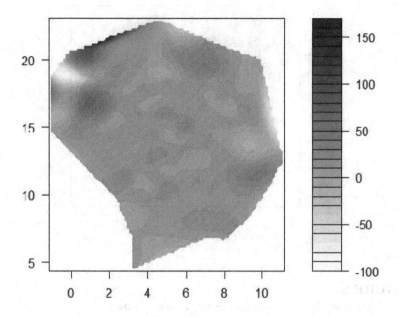

FIGURE 2.2
Filled contour plot of the data in Figure 2.1.

Note that many other commands/packages can be used to help with visualization of this type of data: spplot in sp, ggplot2, ggmap, rgdal, rgeos, and maptools are the most commonly used.

2.2 Universal Kriging

Smoothing packages have embedded algorithms that provide either mathematical or statistical interpolation, usually onto predefined grids. While these packages are quite general, they can be viewed as an added layer, below which a statistical model exists. It is often the case that we want to obtain an optimal interpolation using the most appropriate statistical tools. For 2-D or 3-D surfaces, it is known that universal Kriging has optimal statistical properties

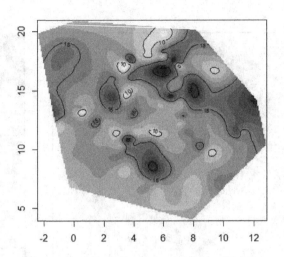

FIGURE 2.3
Multilevel B spline interpolation example using mba.surf.

(Cressie and Wikle, 2011). Because of this, packages that include this form of interpolation are often used. The commonly used GIS package ArcGIS includes functionality for this (Geostatistical Analyst). The Bayesian version of this is Bayesian Kriging. There are several R packages that provide for Bayesian Kriging including spBayes and GeoR. It is not the purpose here to explore Kriging packages in R, but simply to stress that there is a two stage process in visualization. First a statistical model could be used to estimate values within a grid mesh (e.g., Kriging), but secondly a visualization tool will also perform an additional step.

Geostatistical modeling has as its main focus the estimation of continuous spatial fields. As such it is most appropriately used for situations where distance between locations (points) is a fundamental feature. Networks of monitoring sites used for (say) air pollution measurement or geological sampling based on well logs are clear candidates for geostatistical estimation methods. On the other hand health outcomes are usually discrete and based on the experience of individuals. Hence even when residential locations are known for incident cases of disease it is not clear why disease risk should be treated as a continuous process. In the case where an environmental insult is assumed to affect disease risk, then there could be a case made for the use of geostatistical methods. However, in general, this is not the case, and so the discrete nature of risk should be addressed.

2.3   Thematic Mapping

The discrete nature of disease risk coupled with the need to represent spatially aggregated data leads to the consideration of thematic mapping. More specifically, choropleth mapping is a special form of thematic mapping where color gradation is used to depict different levels of the variate of interest. For example, Figure 1.2 displays the SIR thematic map of all respiratory cancers in SC for 1998. This is a choropleth map where shading is used to depict classes of risk.

On R there are a number of resources for producing thematic maps. The main R packages I will examine here are tmap and sp. I will also demonstrate the use of ggplot2 but will not use it extensively. The majority of displays in this work are based on qtm (tmap) and spplot (sp), and I will also use the purpose written R functions fillmap and fillmaps. These latter functions are available from GitHub: https://github.com/carrollrm/fillmap.

2.3.1   Getting map information into R

There are a number of considerations when using R for thematic mapping. First, there is a need to specify a polygon mapping unit. Most functions in R use the object type *spatial polygon object*. This holds information about the vertices of each region to be mapped. There are a number of sources for polygon maps. R itself stores polygon maps for US states and counties within states, Canada, and for a selection of provinces in European countries via the package maps. The basic map function delivers a basic map object including coordinates of polygons and label information. This in turn can be converted to a spatial polygon object. The following code delivers a spatial polygon object in SCpoly2 from the maps library, by using the map2SpatialPolygons function from maptools. Figure 2.4 displays the plot of the polygon object.

```
library(maps)
library(maptools)
SCcounty<-map("county","South Carolina",plot=FALSE,fill=TRUE)
SCpoly2<-map2SpatialPolygons(SCcounty,IDs=SCcounty$names)
plot(SCpoly2)
```

SCmap can now be used as the base polygon object for use in plotting routines.

A common file format that is used extensively by the GIS community is the shapefile (.shp extension). This file is now the commonest format for holding polygon information and many sources provide free downloads for administrative boundaries such as counties, zip codes, postal districts, health service regions etc. In the US, the census bureau holds a large set of shapefiles

FIGURE 2.4
South Carolina counties polygon object.

(Tiger/Line files) for census geographies (such as blocks, block groups, and tracts), as well as zip codes and county boundaries. Other local and national government agencies also often provide shapefiles for their administrative regions. This vector layer shapefile includes a list of all the vertex coordinates of all the polygonal areas on a map. It is accompanied by two other files which hold information about the polygons: a *.dbf and *.shx file. The basic attributes of the polygons, are held in the *.dbf and this can be viewed as a database file on excel. The shapefile for the SC counties is displayed below in the GIS program QuantumGIS (QGIS): Figure 2.5.

On R it is possible to read in vector shapefiles to create polygon objects. Until recently the maptools library hosted a command for this called read-ShapePoly. However this is now deprecated. Instead it is better to use st_read from the sf library. The following code will read the SC county shapefile (SC_county_alphasort.shp) into a spatial polygon object (SCpoly2).

```
library(sf)
SCpoly<-st_read("SC_county_alphasort.shp")
SCpoly2<-as_Spatial(SCpoly)
```

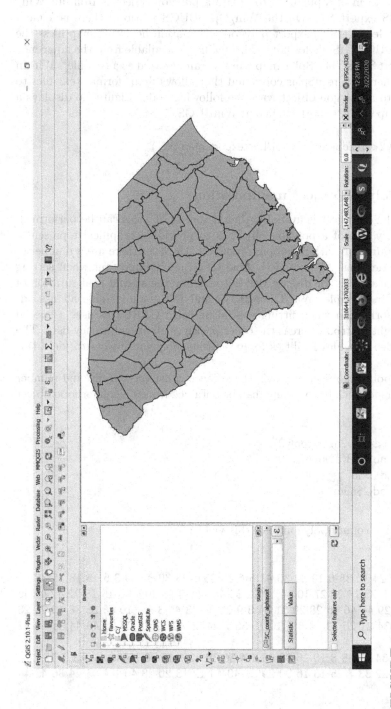

FIGURE 2.5
Quantum GIS display of the SC county polygon shapefile.

The final form of input used to create a polygon object is from an Win/ OpenBUGS export file. On the Win/OpenBUGS platform there is a facility within GeoBUGS to export polygon information held on *.map files (the WinOpenBUGS GIS file format). This facility is available from the adjacency tool: Export Splus. The Splus map window can be saved as a text file. Within maptools there is a readSplus command that allows Splus format text files to be read into a polygon object with the following code. Figure 2.6 displays a screen dump of the export facility in WinBUGS.

```
SCpoly2<-readSplus( "SC_SPLus_export_map.txt" )
```

2.3.2   Polygon object manipulation

Once a polygon object is available then a variety of tasks can be performed. The primary role of a polygon object is to provide a graphical representation in conjunction with a mapping function. However there are a number of other facilities that are provided. One task that is important when setting up Bayesian models for disease risk is to specify the spatial configuration of regions. For example, Win/OpenBUGS and INLA and nimble all use adjacency information when fitting spatial models. CARBayes uses a special binary weight matrix, whereas the other packages all use adjacency lists. The library spdep provides facilities for processing polygon objects to yield this information.

For the polygon object SCpoly2 the adjacency list vector (adj) and number of neighbors for each region (num) can be obtained from a neighborhood object adjnum:

```
adjnum<-poly2nb(SCpoly2)
adj2<-nb2WB(adjnum)
adj<-adj2$adj
num<-adj2$num
```

For the SC county polygon map this yields:

```
adj
 [1] 4 23 24 30 35 6 19 32 38 41 5 6 15 25 1 23 30 37 39 3 6 15 18 25 38
[26] 2 3 5 38 15 25 27 10 14 18 22 38 45 14 32 38 40 43 8 15 18 22 42 44 46
[51] 20 29 44 46 16 28 29 31 34 8 9 21 38 43 45 3 5 7 10 18 25 38 13 21 28
[76] 31 34 21 26 33 34 5 8 10 15 38 2 24 35 41 12 28 29 36 40 44 14 16 17
31
[101] 33 34 43 45 8 10 26 33 45 1 4 30 39 42 1 19 30 35 36 41 3 5 7 15 27
[126] 17 22 33 7 25 13 16 20 29 31 40 43 12 13 20 28 46 1 4 23 24 36 42 44
13
```

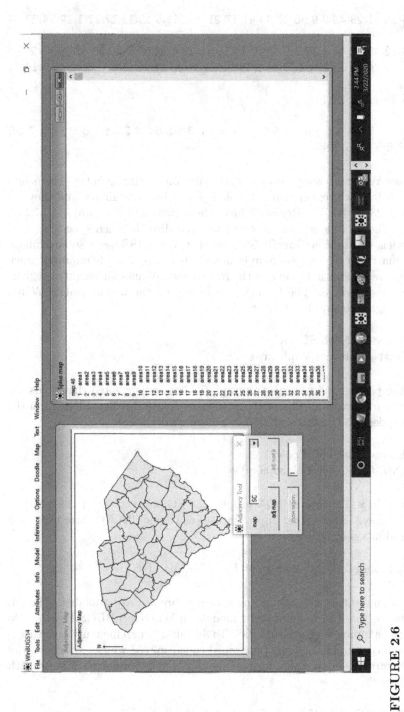

FIGURE 2.6
Screen dump of the Win/OpenBUGS Splus file export facility.

```
[151] 16 21 28 43 2 9 36 38 40 41 17 21 22 26 45 13 16 17 21 1 19 24 20 24
30
[176] 32 40 41 44 4 39 2 5 6 8 9 14 15 18 32 4 23 37 9 20 28 32 36 43 2
[201] 19 24 32 36 11 23 30 44 9 14 21 28 31 40 11 12 20 30 36 42 46 8 14 21
22
[226] 33 11 12 29 44
num
[1] 5 5 4 5 6 4 3 6 5 4 3 4 5 6 7 5 4 5 4 6 8 5 5 6 5 3 2 7 5 7 5 6 5 4 3 7 2 9
[39] 3 6 5 4 6 7 5 4
```

The adj vector is a sparse vector representation specifying in turn the neighbor labels for each region. For example region 1 has 5 neighbors and they are regions 4 23 24 30 35. Region 2 has 5 neighbors and they are 6 19 32 38 41, etc. The vectors adj and num can be used directly when specifying spatial models within Win/OpenBUGS or nimble. For CARBayes a square binary weight matrix of size m (where m is the number of regions) is required. Each row entry represents a region and the row consists of ones for neighbor regions and zeroes elsewhere. The following code sets up the weight matrix W.mat from a neighborhood object W.nb

```
W.nb <- poly2nb(SCpoly2)
W.mat <- nb2mat(W.nb, style="B")
```

For the package INLA, a special graph file has to be created that holds the adjacency information. Ther are two ways to create this file. First, from the library(spdep) the function nb2INLA can be used:

```
adjnum<-poly2nb(SCpoly2)
nb2INLA("SC_poly.txt",adjnum)
```

This code will create a INLA graph file "SC_poly.txt" from the polygon object SCpoly2. The second approach uses the library(INLA) where there is a command inla.geobugs2inla:

```
inla.geobugs2inla(adj,num,graph.file = "SC_poly.txt")
```

This assumes that the adj and num vectors are already available. The adj and num vectors can be used to fit models on Win/OpenBUGS but they do not allow mapping within GeoBUGS. To do this a *.map file must be created. There is a command on library (maptools) called sp2WB which exports spatial polygon objects as an Splus format file that can be imported for use with GeoBUGS:

```
sp2WB(SCpoly2, "SC_SplusRexport.txt")
```

2.4   Mapping Tools

A variety of packages are now available for thematic mapping on R. Here I will review some that are commonly used. First, I will present the R function fillmap which is purpose written and very simple to use.

2.4.1   fillmap

fillmap and its multiple map version fillmaps are purpose written R functions based on the R base plot function.

They use a *spatial polygon object* as the polygon mapping unit. The function is called by:

```
fillmap(map, figtitle, y , n.col, bk="e", cuts,legendtxt="")
```

where map is a polygon object, y is the mappable vector quantity and n.col is the number of colors to use in the legend. Alternatively breakpoints can be specified via either "bk=" or cuts for specific cut points. The fillmap library is available from GitHub: https://github.com/carrollrm/fillmap.

The simplest default call to fillmap with vector y with polygon object SCpoly2 with 5 colors would be

fillmap(SCpoly2, figtitle, y , n.col=5)

A synthetic example could be

```
rand<-rgamma(46,1,1)
fillmap(SCpoly2,"random noise",rand,n.col=5)
```

Figure 2.7 displays the resulting plot.

fillmaps is a multi map extension to fillmap allowing multiple plotting with separate legends. It is available from GitHub by using

```
library(devtools)
install_github("carrollrm/fillmap")
library(fillmap)
```

The full command specification is

```
fillmaps(map, figtitle, y, n.col, bk = "e", cuts, legendtxt = "",
    leg.loc = "bottomright", leg.common = F, lay.m = matrix(1),
    lay.wid = rep.int(1, ncol(lay.m)), leg.cex = 1.5, main.cex = 1.5,
    main.line = -2, lay.hei = rep.int(1, nrow(lay.m)), leg.horiz = F,
    map.lty = 1)
```

random noise

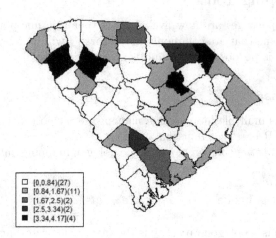

FIGURE 2.7
SC counties: fillmap example using synthetic data.

where map is a spatial polygon object and y can now be a matrix. Separate or common legends can be specified. Figure 1.3 was created using the fillmaps function.

2.4.2   tmap

While fillmaps provides relatively quick and simple processing, tmap is designed to provide sophisticated control for thematic mapping. It is based on the idea of map layers and allows a wide range of layer types and components to be specified. However tmap also includes a very easy to use function called qtm. The full specification of this command I will not demonstrate but a subset appropriate to the mapping task will be provided. An example is

```
qtm(shp,fill=c("obs","expe"),fill.palette="Blues",ncol=2)
```

Here shp is a *spatial polygon dataframe*, which is a spatial polygon object augmented with a matched dataframe of vectors. The plot vectors are "obs," and "expe," the colors used are monochrome blues and 2 columns will be used. shp must be set up with the vectors "obs,"and "expe."

A fuller example is as follows. SCcongen90 is a dataframe consisting of 4 vectors: obs, expe, pov, inc. The following code sets up the spatial polygon dataframe spg.

```
SCcon<-data.frame(SCcongen90)
areaID<-as.character(seq(1:46))
attr<-data.frame(SCcon,row.names=areaID)
spg<-SpatialPolygonsDataFrame(SCpoly2, attr, match.ID = TRUE)
qtm(spg,fill=c("obs","expe","pov","inc"),fill.palette="Blues")
```

Note that the dataframe to merge must have a column matching the ID value in the polygon object. In this case it is simply the order sequence (1:46). Hence attribute row.names has to be added to the SCcon data.frame to give attr. The qtm call delivers a four panel plot of the 4 vectors with separate legends. Some variants of this basic plot can be achieved by extra parameterization and also by using tm_layout. Some examples using 4 years of count data: obsy1,obsy2,obsy3,obsy4 are (1) common legend, (2) legend outside with panel titles, (3) legend positioning:

```
1) qtm(spg,fill=c("obsy1","obsy2","obsy3","obsy4"),fill.palette="Blues",
free.scales=FALSE)

2) labels<-c("count yr 1","count yr 2","count yr 3","count yr 4")
qtm(spg,fill=c("obsy1","obsy2","obsy3","obsy4"),fill.palette="Blues")+tm
_layout(legend.outside=TRUE,title="",panel.labels=labels)

3) qtm(spg,fill=c("obsy1","obsy2","obsy3","obsy4"),fill.palette="Blues")+
tm_layout(title="",panel.labels=labels,legend.position=
c("LEFT","BOTTOM"),legend.height=0.4)
```

Figure 2.8 displays the multi-panel version.
Figure 2.9 displays the separate legends version (3).

2.4.3 spplot

The library sp contains a general purpose spatial plotting function called spplot. This function can be used to create thematic maps. It is especially useful for sequences of vectors (such as time sequences). It also uses a spatial polygon dataframe as its base map type. In the command below spplot uses spg and plots 4 maps.

```
spplot(spg,zcol=c("obsy3","obsy4","obsy1","obsy2"),names.attr=c("count
yr 3","count yr 4","count yr 1","count yr 2"),col.regions=
grey(seq(0.9,0.1,length=40)),layout=c(2,2))
```

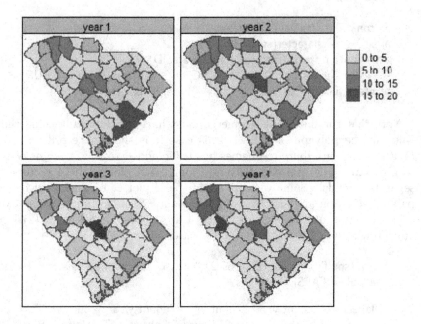

FIGURE 2.8
SC qtm output: multipanel.

Figure 2.10 displays the above spplot command. A disadvantage of spplot is that it does not allow multiple plots with separate legends within each plot, unlike tmap.

2.4.4   ggplot2

The final package that is very popular for mapping is ggplot2. ggplot2 is a very sophisticated graphics package and operates in the form of layers of spatial information. An example of its use is given below. For the spatial polygon dataframe spg the following commands produce a single thematic map of year 1 count.

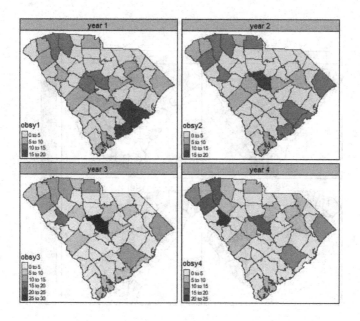

FIGURE 2.9
SC qtm output: separate legends.

```
library(ggplot2)
require("rgdal")
require("plyr")
spg@data$id = rownames(spg@data)
spg.points = fortify(spg, region="id")
spg.df = join(spg.points, spg@data, by="id")
ggplot(spg.df)+
aes(long,lat,group=group,fill=obsy1) +
geom_polygon()
```

The resulting single map plot is displayed in Figure 2.11. While ggplot2 has many flexible features and many dedicated followers, it is more complex and time consuming to set up, even a simple thematic map, and so for this work I have relied on the use of fillmap, tmap, and spplot.

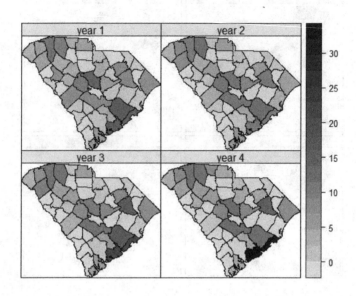

FIGURE 2.10
SC counties spplot with common legend.

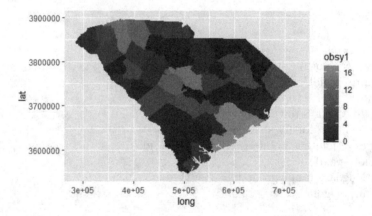

FIGURE 2.11
SC counties year 1 count: ggplot output.

2.5   Chapter Appendix: fillmap R Function Code

fillmap R function code is given below:

```
fillmap<-function(map, figtitle, y , n.col, bk="e", cuts,legendtxt=""){
# map can be a SpatialPolygons object from readSPlus for GEOBUGS
conversion
# eg geobugs<-readSplus("SC_geobugsSPLus.txt")
if(bk=="q"){p <- seq(0,1, length=n.col+1)
 br <- round(quantile(y, probs=p),2)}
if(bk=="e"){br <- round(seq(min(y), max(y), length=n.col+1),2)}
if(bk=="c"){if (length(cuts)!= (n.col+1)) {cat("Cut off and color cate-
gories do not match. ", "\n")
 break} else {br <- cuts} }
# 0: dark 1: light light Current shading ranges from darkest to light gray
white (to distinguish with lakes).
shading<-gray((n.col-1):0/(n.col-1))
y.grp<-findInterval(y, vec=br, rightmost.closed = TRUE, all.inside =
TRUE)
y.shad<-shading[y.grp]
plot(map,col=y.shad,axes=F,main=figtitle)
title(main=figtitle)
br<-round(br, 2)
if (legendtxt=="")
{
cn<-length(y[y>=br[n.col]]) # number of regions in this intervals
 leg.txt<-paste("[",br[n.col],",",br[n.col+1],"]","("",cn,")",sep="")
 for(i in (n.col-1):1){
cn<-length(y[(y>=br[i])&(y<=br[i+1])])
 leg.txt<-append(leg.txt,paste("[",br[i],",",br[i+1],")","("",cn,")",sep=""))
}
 leg.txt<-rev(leg.txt)
 } else {leg.txt<-legendtxt}
legend("bottomleft",legend=leg.txt,fill=shading,cex=0.7,ncol=1,bty="o")
 }
```

3

Bayesian Hierarchical Models

In this section I review some basic concepts in Bayesian hierarchical modeling.

3.1 Likelihood Models

The likelihood for data $\{y_i\}, i = 1, ..., m$, is defined as

$$L(\mathbf{y}|\boldsymbol{\theta}) = \prod_{i=1}^{m} f(y_i|\boldsymbol{\theta}) \tag{3.1}$$

where $\boldsymbol{\theta}$ is a p length vector $\boldsymbol{\theta} : \{\theta_1, \theta_2, ..., \theta_p\}$ and $f(.|.)$ is a probability density (or mass) function. The assumption is made here that the "sample" values of $\mathbf{y}$ given the parameters are independent, and hence it is possible to take the product of individual contributions in (3.1). Hence the data are assumed to be conditionally independent. Note that in many spatial applications the data would not be unconditionally independent and would in fact be correlated. This conditional independence is an important assumption fundamental to many disease mapping applications. The logarithm of the likelihood is also useful in model development and is defined as:

$$l(\mathbf{y}|\boldsymbol{\theta}) = \sum_{i=1}^{m} \log f(y_i|\boldsymbol{\theta}). \tag{3.2}$$

3.2 Prior Distributions

All parameters within Bayesian models are stochastic and are assigned appropriate probability distributions. Hence a single parameter value is simply one possible realization of the possible values of the parameter, the probability of which is defined by the prior distribution. The prior distribution is a distribution assigned to the parameter before seeing the data. Note also that one interpretation of prior distributions are that they provide additional "data"

33

for a problem and so they can be used to improve estimation or identification of parameters. For a single parameter, θ, the prior distribution can be denoted $\mathbf{g}(\theta)$, while for a parameter vector, $\boldsymbol{\theta}$, the joint prior distribution is $\mathbf{g}(\boldsymbol{\theta})$.

3.2.1   Propriety

It is possible that a prior distribution can be *improper*. Impropriety is defined as the condition that integration of the prior distribution of the random variable θ over its range (Ω) is not finite:

$$\int_{\Omega} g(\theta)d\theta = \infty.$$

A prior distribution is improper if its normalizing constant is infinite. While impropriety is a limitation of any prior distribution, it is not necessarily the case that an improper prior will lead to impropriety in the posterior distribution. The posterior distribution can often be proper even with an improper prior specification.

3.2.2   Non-informative priors

Often prior distributions are assumed that do not make strong preferences over values of the variables. These are sometimes known as *vague*, or *reference* or *flat* or *non-informative* prior distributions. Usually, they have a relatively flat form yielding close-to-uniform preference for different values of the variables. This tends to mean that in any posterior analysis (see Section 3.3) that the prior distribution(s) will have little impact compared to the likelihood of the data. *Jeffrey's* priors were developed in an attempt to find such reference priors for given distributions. They are based on the Fisher information matrix. For example, for the binomial data likelihood with common parameter p, then the Jeffrey's prior distribution is $p \sim beta(0.5, 0.5)$. This is a proper prior distribution. However it is not completely non-informative as it has asymptotes close to 0 and 1. Jeffrey's prior for the Poisson data likelihood with common mean θ is given by $g(\theta) \propto \theta^{-\frac{1}{2}}$ which is *improper*. This also is not particularly non-informative. The Jeffrey's prior is locally uniform, however.

Choice of non-informative priors can often be made with some general understanding of the range and behavior of the variable. For example, variance parameters must have prior distributions on the positive real line. Non-informative distributions in this range are often in the gamma, inverse gamma, or uniform families. For example, $\tau \sim gamma(0.001, 0.001)$ will have a small mean (1) but a very large variance (1000) and hence will be relatively flat over a large range. Another specification chosen is $\tau \sim gamma(0.1, 0.1)$ with variance 10 for a more restricted range. On the other hand, a uniform distribution on a large range has been advocated for the standard deviation (Gelman,

2006): $\sqrt{\tau} \sim U(0, 1000)$. For parameters on an infinite range, such as regression parameters, then a distribution centered on zero with a large variance will usually suffice. The zero-mean Gaussian or Laplace distribution could be assumed. For example, a zero-mean Gaussian with variance τ_β, such as

$$\beta \sim N(0, \tau_\beta)$$
$$\tau_\beta = 100000.$$

is typically assumed in applications. The Laplace distribution is favored in machine learning, for large scale Bayesian regression to encourage removal of covariates (Balakrishnan and Madigan, 2006).

Of course sometimes it is important to be informative with prior distributions. Identifiability is an issue relating to the ability to distinguish between parameters within a parametric model (see e.g., Bernardo and Smith, 1994, p. 239). In particular, if a restricted range must be assumed to allow a number of variables to be *identified,* then it may be important to specify distributions that will provide such support. Ultimately if the likelihood has little or no information about the separation of parameters then separation or identification can only come from prior specification. In general, if proper prior distributions are assumed for parameters then they may often be identified in the posterior distribution. However how far they are identified may depend on the assumed variability. An example of identification which arises in disease mapping is where a linear predictor is defined to have two random effect components:

$$\log \theta_i = v_i + u_i,$$

and the components have different normal prior distributions with variances (say, τ_v, τ_u). These variances can have gamma prior distributions such as:

$$\tau_v \sim \text{gamma}(0.001, 0.001)$$
$$\tau_u \sim \text{gamma}(0.1, 0.1).$$

The difference in the variability of the second prior distribution allows there to be some degree of identification. Note that this means that a priori τ_v will be allowed greater variability in the variance of v_i than that found in u_i.

3.3 Posterior Distributions

Prior distributions and likelihood provide two sources of information about any problem. The likelihood informs about the parameter via the data, while the prior distributions inform via prior beliefs or assumptions. When there are large amounts of data, i.e., the sample size is large, the likelihood will contribute more to the relative risk estimation. When the example is data poor then the prior distributions will dominate the analysis.

The product of the likelihood and the prior distributions is called the posterior distribution. This distribution describes the behavior of the parameters after the data are observed and prior assumptions are made. The posterior distribution is defined as:

$$p(\boldsymbol{\theta}|\mathbf{y}) = L(\mathbf{y}|\boldsymbol{\theta})\mathbf{g}(\boldsymbol{\theta})/C \qquad (3.3)$$

$$\text{where } C = \int_p L(\mathbf{y}|\boldsymbol{\theta})\mathbf{g}(\boldsymbol{\theta})d\boldsymbol{\theta}.$$

where $\mathbf{g}(\boldsymbol{\theta})$ is the joint distribution of the $\boldsymbol{\theta}$ vector. Alternatively this distribution can be specified as a proportionality: $p(\boldsymbol{\theta}|\mathbf{y}) \propto L(\mathbf{y}|\boldsymbol{\theta})\mathbf{g}(\boldsymbol{\theta})$.

3.4   Bayesian Hierarchical Modeling

In Bayesian modeling the parameters have distributions. These distributions control the form of the parameters and are specified by the investigator based, usually, on their prior belief concerning their behavior. These distributions are prior distributions denoted by $g(\theta)$.

3.4.1   Hierarchical models

A simple example of a hierarchical model that is commonly found in disease mapping is where the data likelihood is Poisson and there is a common relative risk parameter with a single gamma prior distribution:

$$p(\boldsymbol{\theta}|\mathbf{y}) \propto L(\mathbf{y}|\boldsymbol{\theta})g(\boldsymbol{\theta})$$

where $g(\theta)$ is a gamma distribution with parameters α, β, i.e., gamma(α, β), and $L(y|\theta) = \prod_{i=1}^{m}\{(e_i\theta)^{y_i}\exp(e_i\theta)\}$ bar a constant only dependent in the data. A compact notation for this model is:

$$y_i|\theta \sim Pois(e_i\theta)$$
$$\theta \sim \text{gamma}(\alpha, \beta).$$

In the previous section a simple example of a likelihood and prior distribution was given. In that example the prior distribution for the parameter also had parameters controlling its form. These parameters (α, β) can have assumed values, but more usually an investigator will not have a strong belief in the prior parameters' values. The investigator may want to estimate these parameters from the data. Alternatively and more formally, as parameters within models are regarded as stochastic (and thereby have probability distributions governing their behavior), then these parameters must also have distributions. These distributions are known as hyperprior distributions, and the parameters are known as hyperparameters.

The idea that the values of parameters could arise from distributions is a fundamental feature of Bayesian methodology and leads naturally to the use of models where parameters arise within hierarchies. In the Poisson-gamma example there is a two level hierarchy: θ has a gamma(α, β) distribution at the first level of the hierarchy and α will have a hyperprior distribution (h_α) as will β (h_β), at the second level of the hierarchy. This can be written as:

$$y_i|\theta \sim Pois(e_i\theta)$$
$$\theta|\alpha, \beta \sim \text{gamma}(\alpha, \beta)$$
$$\alpha \sim Exp(v)$$
$$\beta \sim Exp(\rho).$$

3.5 Posterior Inference

When a simple likelihood model is employed, often maximum likelihood is used to provide a point estimate and associated variability for parameters. This is true for simple disease mapping models. When a Bayesian hierarchical model is employed it is no longer possible to provide a simple point estimate for any of the θ_is. This is because the parameter is no longer assumed to be fixed but to arise from a distribution of possible values. Given the observed data, the parameter or parameters of interest will be described by *the posterior distribution of* the θ_is, and hence this distribution must be found and examined. It is possible to examine the expected value (mean) or the mode of the posterior distribution to give a point estimate for a parameter or parameters: e.g., for a single parameter θ, say, then $E(\theta|y) = \int \theta \, p(\theta|y)d\theta$, or $\arg\max_\theta p(\theta|y)$. Just as the maximum likelihood estimate is the mode of the likelihood, then the maximum a posteriori estimate is that value of the parameter or parameters at the mode of the posterior distribution. More commonly the expected value of the parameter or parameters is used. This is known as the posterior mean (or Bayes estimate). For simple unimodal symmetrical distributions, the modal and mean estimates coincide.

For some simple posterior distributions it is possible to find the exact form of the posterior distribution and to find explicit forms for the posterior mean or mode. However, it is commonly the case that for reasonably realistic models within disease mapping, it is not possible to obtain a closed form for the posterior distribution. Hence it is often not possible to derive simple estimators for parameters such as the relative risk. In this situation resort must be made to posterior approximation either via numerical approximation or by posterior sampling. In the latter case, simulation methods are used to obtain samples from the posterior distribution which then can be summarized to yield estimates of relevant quantities. In the next section the use of sampling algorithms for this purpose is discussed.

3.5.1   A Bernoulli and binomial example

Another example of a model hierarchy that arises commonly is the small area health data is where a finite population exists within an area and within that population binary outcomes are observed. A fuller discussion of these models is given in Section 6.1.3. In the case event example, define the case events as $s_i : i = 1, ..., m$ and the control events as $s_i : i = m + 1, ..., N$ where $N = m + n$ the total number of events in the study area. Associated with each location is a binary variable (y_i) which labels the event either as a case ($y_i = 1$) or a control ($y_i = 0$). A conditional Bernoulli model is assumed for the binary outcome where p_i is the probability of an individual being a case, given the location of the individual. Hence we can specify that $y_i \sim Bern(p_i)$. Here the probability will usually have either a prior distribution associated with it, or will be linked to other parameters and covariate or random effects, possibly via a linear predictor. Assume that a logistic link is appropriate for the probability and that two covariates are available for the individual: x_1 : age, x_2 : exposure level (of a health hazard). Hence,

$$p_i = \frac{\exp(\alpha_0 + \alpha_1 x_{1i} + \alpha_2 x_{2i})}{1 + \exp(\alpha_0 + \alpha_1 x_{1i} + \alpha_2 x_{2i})}$$

is a valid logistic model for this data with three parameters ($\alpha_0, \alpha_1, \alpha_2$). Assume that the regression parameters will have independent zero-mean Gaussian prior distributions. The hierarchical model is specified in this case as:

$$y_i | p_i \sim Bern(p_i)$$
$$logit(p_i) = \mathbf{x}_i' \boldsymbol{\alpha}$$
$$\alpha_j | \tau_j \sim N(0, \tau_j)$$
$$\tau_j \sim gamma(a, b).$$

In this case, x_i' is the i th row of the design matrix (including an intercept term), $\boldsymbol{\alpha}$ is the (3×1) parameter vector, τ_j is the variance for the j th parameter, and a, and b are fixed scale and shape parameters.

3.5.2   Random effects with a binomial example

In the binomial case we could have a collection of small areas within which we observe events. Define the number of small areas as m and the total small area population as n_i. Within the population of each area individuals have a binary label which denotes the case status of the individual. The number of cases are denoted as y_i and it is often assumed that the cases follow an independent binomial distribution, conditional on the probability that an individual is a case, defined as p_i: $y_i \sim bin(p_i, n_i)$.

The likelihood is given by $L(y_i|p_i, n_i) = \prod_{i=1}^{m} \binom{n_i}{y_i} p_i^{y_i} (1-p_i)^{(n_i-y_i)}$. Here the probability will usually have either a prior distribution associated with it, or will be linked to other parameters and predictor or *random effects*, possibly via a linear predictor such as $\text{logit}(p_i) = x_i'\alpha + z_i'\gamma$.

Random effects are commonly introduced in observational studies. These effects account for unobserved confounding in the outcome of interest. Either there are known confounders that are not observed in the study or there could be unknown confounders. In the latter case these could be thought to exist in most observational studies as we have imperfect knowledge of the etiology of disease. Essentially the random effects are absorbing extra noise unaccounted for in the model specified.

In this general case, the z_i' are a vector of individual level random effects and the γ is a unit vector. Assume that a logistic link is appropriate for the probability and that a random effect at the individual level is to be included: v_i. Hence,

$$p_i = \frac{\exp(\alpha_0 + v_i)}{1 + \exp(\alpha_0 + v_i)}$$

would represent a basic model with intercept to capture the overall rate and prior distribution for the intercept and the random effect could be assumed to be $\alpha_0 \sim N(0, \tau_{\alpha_0})$, and $v_i \sim N(0, \tau_v)$. The hyperprior distribution for the variance parameters could be a distribution on the positive real line such as the gamma, inverse gamma, or uniform. The uniform distribution has been proposed for the standard deviation ($\sqrt{\tau_*}$) by Gelman (2006). Here for illustration, I define a gamma distribution:

$$y_i \sim bin(p_i, n_i)$$
$$logit(p_i) = \alpha_0 + v_i$$
$$\alpha_0 \sim N(0, \tau_{\alpha_0})$$
$$v_i \sim N(0, \tau_v)$$
$$\tau_{\alpha_0} \sim \text{gamma}(\psi_1, \psi_2)$$
$$\tau_v \sim \text{gamma}(\phi_1, \phi_2)$$

An alternative approach to the Bernoulli or binomial distribution at the second level of the hierarchy is to assume a distribution directly for the case probability p_i. This might be appropriate when limited information about p_i is available. This is akin to the assumption of a gamma distribution as prior distribution for the Poisson relative risk parameter. Here one choice for the prior distribution could be a beta distribution:

$$p_i \sim beta(\alpha_1, \alpha_2).$$

In general, the parameters α_1 and α_2 could be assigned hyperprior distributions on the positive real line, such as gamma or exponential. However if a uniform prior distribution for p_i is favored then $\alpha_1 = \alpha_2 = 1$ can be chosen.

4

Computation

4.1 Posterior Sampling

Once a posterior distribution has been derived, from the product of likelihood and prior distributions, it is important to assess how the form of the posterior distribution is to be evaluated. If single summary measures are needed then it is sometimes possible to obtain these directly from the posterior distribution either by direct maximization (mode: maximum a posteriori estimation) or analytically in simple cases (mean or variance for example)(see Section 3.3). If a variety of features of the posterior distribution are to be examined then often it will be important to be able to access the distribution via posterior sampling. Posterior sampling is a fundamental tool for exploration of posterior distributions and can provide a wide range of information about their form. Define a posterior distribution for data $\mathbf{y}$ and parameter vector $\boldsymbol{\theta}$ as $p(\boldsymbol{\theta}|\mathbf{y})$. We wish to represent features of this distribution by taking a sample from $p(\boldsymbol{\theta}|\mathbf{y})$. The sample can be used to estimate a variety of posterior quantities of interest. Define the sample size as m_p.

For analytically tractable posterior distributions may be available to directly simulate the distribution. For example the Gamma-Poisson model with α, β known, in Section 3.4.1, leads to the gamma posterior distribution: $\theta_i \sim gamma(y_i + \alpha, e_i + \beta)$. This can either be simulated directly (on R: rgamma) or sample estimation can be avoided by direct computation from known formulas. For example, in this instance, the moments of a Gamma distribution are known: $E(\theta_i) = (y_i + \alpha)/(e_i + \beta)$ etc.

Define the sample values generated as: θ_{ij}^*, $j = 1, ..., m_p$. As long as a sample of reasonable size has been taken then it is possible to approximate the various functionals of the posterior distribution from these sample values. For example, an estimate of the posterior mean would be $\widehat{E}(\theta_i) = \widehat{\theta}_i = \sum_{j=1}^{m_p} \theta_{ij}^*/m_p$, while the posterior variance could be estimated as $\widehat{var}(\theta_i) = \frac{1}{m_p-1}\sum_{j=1}^{m_p}(\theta_{ij}^* - \widehat{\theta}_i)^2$, the sample variance. In general, any real function of the j th parameter $\gamma_j = t(\theta_j)$ can also be estimated in this way. For example, the mean of γ_j

is given by $\widehat{E}(\gamma_j) = \widehat{\gamma}_j = \sum_{j=1}^{m_p} t(\theta_{ij}^*)/m_p$. Note that credibility intervals can also be found for parameters by estimating the respective sample quantiles. For example if $m_p = 1000$ then 25 th and 975 th largest values would yield an equal tail 95% credible interval for γ_j. The median is also available as the 50% percentile of the sample, as are other percentiles.

The empirical distribution of the sample values can also provide an estimate of the marginal posterior density of θ_i. Denote this density as $\pi(\theta_i)$. A smoothed estimate of this marginal density can be obtained from the histogram of sample values of θ_i. Improved estimators can be obtained by using conditional distributions. A Monte Carlo estimator of $\pi(\theta_i)$ is given by

$$\widehat{\pi}(\theta_i) = \frac{1}{n} \sum_{j=1}^{n} \pi(\theta_i | \theta_{j,-i})$$

where the $\theta_{j,-i}\ j = 1, ..., n$ are a sample from the marginal distribution $\pi(\theta_{-i})$

Often m_p is chosen to be ≥ 500, more often 1000 or 10,000. If computation is not expensive then large samples such as these are easily obtained. The larger the sample size the closer the posterior sample estimate of the functional will be.

Generally, the complete sample output from the distribution is used to estimate functionals. This is certainly true in the case when independent sample values are available (such as when the distribution is analytically tractable and can be sampled from directly, such as in the gamma-Poisson case). In other cases, where iterative sampling must be used, it is sometimes necessary to subsample the output sample. In the next section, this is discussed more fully.

4.2   Markov Chain Monte Carlo Methods

Often in disease mapping, realistic models for maps have two or more levels and the resulting complexity of the posterior distribution of the parameters requires the use of sampling algorithms. In addition, the flexible modelling of disease could require switching between a variety of relatively complex models. In this case, it is convenient to have an efficient and flexible posterior sampling method which could be applied across a variety of models. Efficient algorithms for this purpose were developed within the fields of physics and image processing to handle large scale problems in estimation. In the late 1980s and early 1990s these methods were developed further particularly for dealing with Bayesian posterior sampling for more general classes of problems (Gilks et al., 1993, 1996). Now posterior sampling is commonplace and a variety of packages (including WinBUGS, its descendent OpenBUGS, R, with MCMCpack, CARBayes and R-NIMBLE) have incorporated these methods. In chapter 4 we will explore further these packages in disease mapping applications. For general reviews of this area the reader is referred to Cassella and George (1992), Robert and Casella (2005), Gamerman and Lopes (2006), and Brooks et al. (2011). Markov chain Monte Carlo (McMC) methods are a set of methods which use iterative simulation of parameter values within a Markov chain. The convergence of this chain to a stationary distribution, which is assumed to be the posterior distribution, must be assessed.

Prior distributions for the p components of $\boldsymbol{\theta}$ are usually defined independently, as $g_i(\theta_i)$ for $i = 1, ..., p$. The posterior distribution of $\boldsymbol{\theta}$ and $\mathbf{y}$ is defined as:

$$P(\boldsymbol{\theta}|\mathbf{y}) \propto L(\mathbf{y}|\boldsymbol{\theta}) \prod_i g_i(\theta_i). \tag{4.1}$$

The aim is to generate a sample from the posterior distribution $P(\boldsymbol{\theta}|\mathbf{y})$. Suppose we can construct a Markov chain with state space $\boldsymbol{\theta}_c$, where $\boldsymbol{\theta} \in \boldsymbol{\theta}_c \subset \Re^k$. The chain is constructed so that the equilibrium distribution is $P(\boldsymbol{\theta}|\mathbf{y})$, and the chain should be easy to simulate from. If the chain is run over a long period, then it should be possible to reconstruct features of $P(\boldsymbol{\theta}|\mathbf{y})$ from the realized chain values. This forms the basis of the McMC method, and algorithms are required for the construction of such chains. A selection of literature on this area is found in Ripley (1987), Gelman and Rubin (1992), Smith and Roberts (1993), Besag and Green (1993), Smith and Gelfand (1992), Tanner (1996), Chen et al. (2000), Robert and Casella (2005). McMC methods are now commonly applied in spatial statistical modeling: see for example Cressie and Wikle (2011), Gelfand et al. (2010), and Banerjee et al. (2014).

The basic algorithms used for this construction are:

1. the Metropolis and its extension Metropolis-Hastings algorithm

2. the Gibbs Sampler algorithm

In all McMC algorithms, it is important to be able to construct the correct *transition probabilities* for a chain which has $P(\boldsymbol{\theta}|\mathbf{y})$ as its equilibrium distribution. A Markov chain consisting of $\boldsymbol{\theta}^1, \boldsymbol{\theta}^2, ...\boldsymbol{\theta}^t$ with state space Θ and equilibrium distribution $P(\boldsymbol{\theta}|\mathbf{y})$ has transitions defined as follows.

Define $q(\boldsymbol{\theta}, \boldsymbol{\theta}')$ as a transition probability function, such that, if $\boldsymbol{\theta}^t = \boldsymbol{\theta}$, the vector $\boldsymbol{\theta}^t$ drawn from $q(\boldsymbol{\theta}, \boldsymbol{\theta}')$ is regarded as a proposed possible value for $\boldsymbol{\theta}^{t+1}$.

4.2.1 Metropolis updates

In this case choose a symmetric proposal $q(\boldsymbol{\theta}, \boldsymbol{\theta}')$ and define the transition probability as

$$p(\boldsymbol{\theta}, \boldsymbol{\theta}') = \begin{cases} \alpha(\boldsymbol{\theta}, \boldsymbol{\theta}')q(\boldsymbol{\theta}, \boldsymbol{\theta}') & \text{if } \boldsymbol{\theta}' \neq \boldsymbol{\theta} \\ 1 - \sum_{\theta''} q(\boldsymbol{\theta}, \boldsymbol{\theta}'')\alpha(\boldsymbol{\theta}, \boldsymbol{\theta}'') & \text{if } \boldsymbol{\theta}' = \boldsymbol{\theta} \end{cases}$$

where $\alpha(\boldsymbol{\theta}, \boldsymbol{\theta}') = \min\left\{1, \frac{P(\boldsymbol{\theta}'|\mathbf{y})}{P(\boldsymbol{\theta}|\mathbf{y})}\right\}$

In this algorithm a proposal is generated from $q(\boldsymbol{\theta}, \boldsymbol{\theta}')$ and is accepted with probability $\alpha(\boldsymbol{\theta}, \boldsymbol{\theta}')$. The acceptance probability is a simple function of the ratio of posterior distributions as a function of the ratio of posterior distributions as a function of $\boldsymbol{\theta}$ values. The proposal function $q(\boldsymbol{\theta}, \boldsymbol{\theta}')$ can be defined to have a variety of forms but must be an irreducible and aperiodic transition function. Specific choices of $q(\boldsymbol{\theta}, \boldsymbol{\theta}')$ lead to specific algorithms.

4.2.2 Metropolis-Hastings updates

In this extension to the Metropolis algorithm the proposal function is not confined to symmetry and

$$\alpha(\boldsymbol{\theta}, \boldsymbol{\theta}') = \min\left\{1, \frac{P(\boldsymbol{\theta}'|\mathbf{y})q(\boldsymbol{\theta}', \boldsymbol{\theta})}{P(\boldsymbol{\theta}|\mathbf{y})q(\boldsymbol{\theta}, \boldsymbol{\theta}')}\right\}.$$

Some special cases of chains are found when $q(\boldsymbol{\theta}, \boldsymbol{\theta}')$ has special forms. For example, if $q(\boldsymbol{\theta}, \boldsymbol{\theta}') = q(\boldsymbol{\theta}', \boldsymbol{\theta})$ then the original Metropolis method arises and further, with $q(\boldsymbol{\theta}, \boldsymbol{\theta}') = q(\boldsymbol{\theta}')$, (i.e., when no dependence on the previous value is assumed) then

$$\alpha(\boldsymbol{\theta}, \boldsymbol{\theta}') = \min\left\{1, \frac{w(\boldsymbol{\theta}')}{w(\boldsymbol{\theta})}\right\}$$

where $w(\boldsymbol{\theta}) = P(\boldsymbol{\theta}|\mathbf{y})/q(\boldsymbol{\theta})$ and $w(.)$ are importance weights. One simple example of the method is $q(\boldsymbol{\theta}') \sim \text{Uniform}(\boldsymbol{\theta}_a, \boldsymbol{\theta}_b)$ and $g_i(\theta_i) \sim \text{Uniform}(\theta_{ia}, \theta_{ib})$ $\forall i$, this leads to an acceptance criterion based on a likelihood ratio. Hence the original Metropolis algorithm with uniform proposals and prior distributions leads to a stochastic exploration of a likelihood surface. This, in effect, leads to the use of prior distributions as proposals. However, in general, when the $g_i(\theta_i)$ are not uniform this leads to inefficient sampling. The definition of $q(\boldsymbol{\theta}, \boldsymbol{\theta}')$ can be quite general in this algorithm and, in addition, the posterior distribution only appears within a ratio as a function of $\boldsymbol{\theta}$ and $\boldsymbol{\theta}'$. Hence, the distribution is only required to be known up to proportionality.

An example of a commonly used MH update for a continuous parameter is the random walk Metropolis (RWM): this essentially assumes a Gaussian proposal based on a mean which is the previous parameter value. Hence the update is

$$\theta' < -N(\theta, \tau_\theta) \tag{4.2}$$

where τ_θ defines the variance of the proposal.

4.2.3 Gibbs updates

The Gibbs Sampler has gained considerable popularity, particularly in applications in medicine, where hierarchical Bayesian models are commonly applied (see, e.g., Gilks et al., 1993). This popularity is mirrored in the availability of software which allows its application in a variety of problems (e.g., Win/OpenBUGS, MLwiN, JAGS, MCMCpack, NIMBLE on R). This sampler is a special case of the Metropolis-Hastings algorithm where the proposal is generated from the conditional distribution of θ_i given all other $\boldsymbol{\theta}$'s, and the resulting proposal value is accepted with probability 1.

More formally, define

$$q(\theta_j, \theta_j') = \begin{cases} p(\theta_j^* | \theta_{-j}^{t-1}) & if \ \theta_{-j}^* = \theta_{-j}^{t-1} \\ 0 & \text{otherwise} \end{cases}$$

where $p(\theta_j^* | \theta_{-j}^{t-1})$ is the conditional distribution of θ_j given all other $\boldsymbol{\theta}$ values (θ_{-j}) at time $t-1$. Using this definition it is straightforward to show that

$$\frac{q(\boldsymbol{\theta}, \boldsymbol{\theta}')}{q(\boldsymbol{\theta}', \boldsymbol{\theta})} = \frac{P(\boldsymbol{\theta}'|\mathbf{y})}{P(\boldsymbol{\theta}|\mathbf{y})}$$

and hence $\alpha(\boldsymbol{\theta}, \boldsymbol{\theta}') = 1$.

4.2.4   M-H versus Gibbs algorithms

There are advantages and disadvantages to M-H and Gibbs methods. The Gibbs Sampler provides a *single* new value for each $\boldsymbol{\theta}$ at each iteration, but requires the evaluation of a conditional distribution. On the other hand the M-H step does not require evaluation of a conditional distribution but does not guarantee the acceptance of a new value. In addition, block updates of parameters are available in M-H, but not usually in Gibbs steps (unless joint conditional distributions are available). If conditional distributions are difficult to obtain or computationally expensive, then M-H can be used and is usually available.

In summary, the Gibbs Sampler may provide faster convergence of the chain if the computation of the conditional distributions at each iteration are not time consuming. The M-H step will usually be faster at each iteration, but will not necessarily guarantee exploration. In straightforward hierarchical models where conditional distributions are easily obtained and simulated from, then the Gibbs Sampler is likely to be favored. In more complex problems, such as many arising in spatial statistics, resort may be required to the M-H algorithm.

4.2.5   Special methods

Alternative methods exist for posterior sampling when the basic Gibbs or M-H updates are not feasible or appropriate. For example, if the range of the parameters are restricted then slice sampling can be used (Robert and Casella, 2005 , Chapter 7; Neal, 2003). When exact conditional distributions are not available but the posterior is log-concave then adaptive rejection sampling algorithms can be used. The most general of these algorithms (ARS algorithm;

Robert and Casella, 2005, p.57-59) has wide applicability for continuous distributions, although may not be efficient for specific cases. Block updating can also be used to effect in some situations. When generalized linear model components are included then block updating of the covariate parameters can be effected via multivariate updating.

A variant known as Metropolis adjusted Langevin Algorithm (MALA) has been proposed and includes a gradient search (Langevin dynamic) before a MH update. This can lead to more efficient searching of the posterior space. This is a special case of Hamiltonian Monte Carlo (HMC; Neal, 2011). The R packages GeoRglm and CARBayes use this approach. In addition, the programming package STAN also implements this HMC approach (NUTS: No-U-Turn sampler, Hoffman and Gelman (2014)). In general, the acceptance rate of these samplers is higher than standard MH samplers.

4.2.6   Convergence

McMC methods require the use of diagnostics to assess whether the iterative simulations have reached the equilibrium distribution of the Markov chain. Sampled chains require to be run for an initial burn-in period until they can be assumed to provide approximately correct samples from the posterior distribution of interest. This burn-in period can vary considerably between different problems. In addition, it is important to ensure that the chain manages to explore the parameter space properly so that the sampler does not "stick" in local maxima of the surface of the distribution. Hence, it is crucial to ensure that a burn-in period is adequate for the problem considered. Judging convergence has been the subject of much debate and can still be regarded as art rather than science: a qualitative judgement has to be made at some stage as to whether the burn-in period is long enough.

There are a wide variety of methods now available to assess convergence of chains within McMC. Robert and Casella (2005) and Liu (2001) provide reviews. Visual inspection of chains is often a first pass informal metric. Figure 4.1   displays the trace plot for 2 chains (red and blue) for a model parameter alpha. The alpha parameter appears to be varying around a stationary level (mean alpha of 1.21 in converged sample) and the plots overwrite each other which suggests that the chains are mixing well. Hence this might be suggestive of convergence, but convergence must be checked formally as well. In addition it is possible for one parameter to converge, but some parameters to remain unconverged. This may possibly be due to identifiability issues in the model, but the sampler may also require longer run times to achieve the converged state. Usually it would be expected that all parameters pass convergence checks to declare convergence. Initial checking is often done based on the deviance measure $(D = -2l(\mathbf{y}|\boldsymbol{\theta}))$ which provides a measure of overall goodness of fit and so could represent overall model convergence.

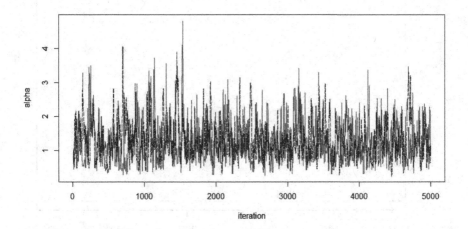

FIGURE 4.1
Trace plot of two chains run over 5000 iterations.

Formal methods are largely based on checking the distributional properties of samples from the chains. In general define an output stream for a parameter vector $\boldsymbol{\theta}$ as $\left\{ \boldsymbol{\theta}^1, \boldsymbol{\theta}^2, ...\boldsymbol{\theta}^m, \boldsymbol{\theta}^{m+1}....\boldsymbol{\theta}^{m+m_p} \right\}$. Here the mth value is the end of the burn-in period and a (converged) sample of size m_p is taken. Hence the converged sample is $\left\{ \boldsymbol{\theta}^{m+1}....\boldsymbol{\theta}^{m+m_p} \right\}$. Define a function of the output stream as $\gamma = t(\boldsymbol{\theta})$ so that $\gamma^1 = t(\boldsymbol{\theta}^1)$.

4.2.6.1 Single chain methods

First, global methods for assessing convergence have been proposed which involve monitoring functions of the posterior output at each iteration. Globally this output could be the log posterior value (log $p(\widehat{\boldsymbol{\theta}}|\mathbf{y})$ where $\widehat{\boldsymbol{\theta}}$ are the estimated parameters at a given iteration), or the deviance of the model $(-2[l(y|\widehat{\boldsymbol{\theta}}) - l(y|\widehat{\boldsymbol{\theta}}_{ref})]$ where $\widehat{\boldsymbol{\theta}}_{ref}$ is a saturated or other reference model estimate). (In Win/OpenBUGS the deviance is assumed to be $-2l(y|\widehat{\boldsymbol{\theta}})$). These methods look for stabilization of the probability value. This value forms a time series, and special cusum methods have been proposed (Yu and Mykland, 1998). This approach emphasizes the overall convergence of the chain rather than individual parameter convergence. Two basic statistical tools that can be used to check sequences of output, have been proposed by Geweke (1992) and Yu and Mykland (1998). For the Geweke statistic, the sequence of output is broken up into two segments following a burn - in of m length. The first and last segments of length n_b and n_a respectively are defined. Averages of

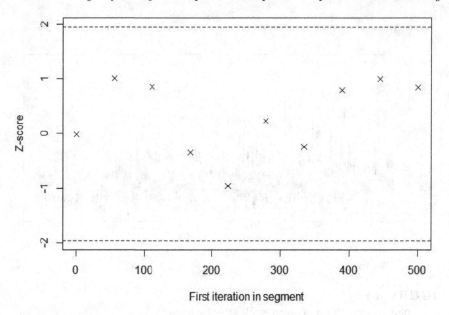

FIGURE 4.2
Geweke diagnostic plot for the deviance measure from a single chain sample of 500 iterations. Dotted lines denote acceptable limits for convergence.

the first and last segments of output are obtained:

$$\overline{\gamma}_b = \frac{1}{n_b} \sum_{j=m+1}^{m+n_b} \gamma^j$$

$$\overline{\gamma}_a = \frac{1}{n_a} \sum_{j=m+m_p-n_a+1}^{m+m_p} \gamma^j.$$

As m_p gets large then the statistic

$$G = \frac{\overline{\gamma}_a - \overline{\gamma}_b}{\sqrt{\widehat{var}(\gamma_a) + \widehat{var}(\gamma_b)}} \rightarrow N(0,1) \text{ in distribution,}$$

where $\widehat{var}(\gamma_a), \widehat{var}(\gamma_b)$ are empirical variance estimates. Usually it is assumed that $n_b = 0.1n$ and $n_a = 0.5n$, where n is the sample taken. Note that we can set $\gamma^j = -2l(y|\boldsymbol{\theta}^j)$ or $\gamma^j = \log p(\boldsymbol{\theta}^j|\mathbf{y})$ and so the deviance or log posterior can be monitored as an overall measure. This test is available on R in the CODA package (**geweke.diag**). CODA is the R package for convergence diagnostics and contains a variety of test diagnostics for McMC convergence. Figure 4.2 displays the Geweke diagnostics for a deviance measure from a single chain.

It is also available in the R package CARBayes as the main single chain diagnostic criterion. The second test for single sequences was proposed by Yu and Mykland (1998) and later modified by Brooks (1998). For a post-convergence sequence of length m_p an average is computed

$$\widehat{\mu} = \frac{1}{m_p} \sum_{j=m+1}^{m+m_p} \gamma^j.$$

This average is used within a cusum calculation by defining a cusum of the sequence:

$$\widehat{S}_t = \sum_{j=m+1}^{t} [\gamma^j - \widehat{\mu}] \quad \text{for } t = m+1, .., m+m_p.$$

In the original proposal, a plot of $\widehat{S}_t$ against t was proposed. The interpretation of the plot relies on the identification of the hairiness or spikeyness of the cusum: a smooth cusum suggesting under-exploration of the posterior distribution, while a spikey plot represents rapid mixing. Brooks (1998) further quantified this approach by deriving a statistic that measures the spikeyness of $\widehat{S}_t$.

Second, graphical methods have been proposed which allow the comparison of the whole distribution of successive samples. Quantile-quantile plots of successive lengths of single variable output from the sampler can be used for this purpose. Figure 4.3 displays an example of such a plot. On R, with vectors out1 and out2 this can be created via commands:

```
plot(sort(out1),sort(out2),xlab="output stream 1",ylab="output stream 2")
lines(x,y)
cor(sort(out1),sort(out2))
```

Further assessment of the degree of equality can be made via use of a correlation test. The Pearson correlation coefficient between the sorted sequences can be examined and compared to special tables of critical values. This adds some formality to the relatively arbitrary nature of visual inspection.

4.2.6.2 Multi-chain methods

Single chain methods can, of course, be applied to each of a multiple of chains. In addition, there are methods that can only be used for multiple chains. The Brooks-Gelman-Rubin (BGR) statistic was proposed as a method for assessing the convergence of multiple chains via the comparison of summary measures across chains (Gelman and Rubin, 1992, Brooks and Gelman, 1998, Robert and Casella, 2005, Chapter 8, Gelman and Shirley, 2011).

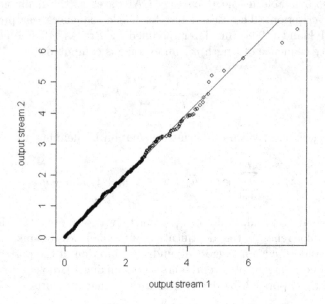

FIGURE 4.3
Quantile-quantile plot of two sequences of 1000 length of converged sample output from a gamma posterior distribution with parameters $\alpha = 1, \beta = 1$. The equality line is marked.

   This statistic is based on between and within chain variances. For the univariate case we have p chains and a sample of size n and a sample value of γ_i^j $j = 1, ..., n;$ $i = 1, ..., p$. Denote the average over the sample for the i th chain as $\overline{\gamma}_i = \frac{1}{n} \sum_{j=1}^{n} \gamma_i^j$ and the overall average as $\overline{\gamma}_. = \frac{1}{p} \sum_{i=1}^{p} \overline{\gamma}_i$ and the variance of the i th chain is $\tau_i^2 = \frac{1}{n-1} \sum_{j=1}^{n} (\gamma_i^j - \overline{\gamma}_i)^2$. Then the between- and within - sequence variances are

$$B = \frac{n}{p-1} \sum_{i=1}^{p} (\overline{\gamma}_i - \overline{\gamma}_.)^2$$

$$W = \frac{1}{p} \sum_{i=1}^{p} \tau_i^2.$$

The marginal posterior variance of the γ is estimated as $\frac{n-1}{n} W + \frac{1}{n} B$ and this is unbiased asymptotically ($n \to \infty$) . Monitoring the statistic

$$R = \sqrt{\frac{n-1}{n} + \frac{1}{n}\frac{B}{W}}$$

for convergence to 1 is recommended. If the R for all parameters and functions of parameters are between 1.0 and 1.1 (Gelman et al., 2004) this is acceptable for most studies. Note that this depends on the sample size taken and closeness will be more easily achieved for a large post - convergence m_p. Brooks and Gelman (1998) extended this diagnostic to a multiparameter situation. On R the statistic is available in the coda package as gelman.diag. On Win/OpenBUGS the Brooks-Gelman-Rubin (BGR) statistic is available in the Sample Monitor Tool. On Win/OpenBUGS, the width of the central 80% interval of the pooled runs and the average width of the 80% intervals within the individual runs are color-coded (green, blue), and their ratio R is red - for plotting purposes the pooled and within interval widths are normalized to have an overall maximum of one. On Win/OpenBUGS the statistics are calculated in bins of length 50. R would generally be expected to be greater than 1 if the starting values are suitably over-dispersed. Brooks and Gelman (1998) emphasize that one should be concerned both with convergence of R to 1, and with convergence of both the pooled and within interval widths to stability. One caveat should be mentioned concerning the use of between and within chain diagnostics. If the posterior distribution being approximated were to be highly multimodal, which could be the case in many mixture and spatial problems, then the variability across chains could be large even when close to the posterior distribution and it could be that very large bins would need to be used for computation. Hence the default WinBUGS bins may be inappropriate.

Figure 4.4 displays a BGR plot from the coda package for parameters alpha and beta for a model with a gamma prior distribution (i.e., $ga(alpha, beta)$)run on nimble (see chapter 8). After an initial spike, which can commonly occur due to edge effects, the plots settle down to close to 1.0 for most iterations

There is some debate about whether it is useful to run one long chain as opposed to multiple chains with different start points. The advantage of multiple chains is that they provide evidence for the robustness of convergence across different subspaces. However, as long as a single chain samples the parameter space adequately, then these have benefits. One major advantage of single chains is that they save computation time, and so for chains requiring long runs to get convergence single chains may be preferred. The package CARBayes employs single chains and provide Geweke diagnostics automatically (see chapter 9). Also multi chain diagnostics can also be applied to single chains by splitting the chain into segments. The reader is referred to Robert and Casella (2005), chapter 8 and Gelman and Shirley (2011) for a thorough discussion of diagnostics and their use.

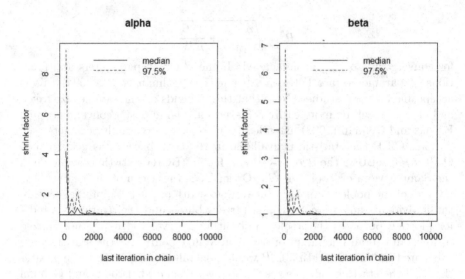

FIGURE 4.4
Gelman-Rubin plots using the gelman.plot function in the coda package.

4.2.7   Subsampling and thinning

McMC samplers often produce correlated samples of parameters. That is, a parameter value γ_i^j is likely to be similar to γ_i^{j-1}. This is likely to be true if successful proposals are based on proposal distributions with small variances, or where acceptances are localized to small areas of the posterior surface. In the former case, it may be that only small subsections of the posterior surface are being explored and so the sampler will not reach equilibrium for some time. Hence there may be an issue of lack of convergence when this occurs. The latter case could arise when a very spikey likelihood dominates. In themselves these correlated samples do not create problems for subsequent use of output streams, unless the sample size is very small (m_p small), or convergence has not been reached. Summary statistics could be affected by such autocorrelation. While measures of central tendency may not be much affected, the variance and other spread measures could be downward biased due to the (positive) autocorrelation in the stream. One possible remedy for this correlation is to take subsamples of the output. The simplest approach to this is to *thin* the stream by taking systematic samples at every k th iteration. By lengthening the gap between sampled units, then the more likely the correlation will be reduced or eliminated.

As a diagnostic tool, an autocorrelation function (ACF) can be viewed, and this displays the correlation between the current iteration and lagged previous iterations. The R libraries coda and smfsb include facilities for checking ACFs.

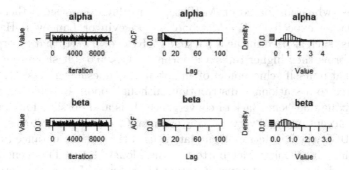

FIGURE 4.5
mcmcSummary output for alpha and beta parameters.

On coda the acfplot function can be used. Package smfsb was developed for stochastic modeling for systems biology and includes a function mcmcSummary that provides trace plot, acf and marginal density of the sampled parameters. Figure 4.5 displays the trace plots, acfs and densities for the sampled alpha and beta parameters. The acfs suggest a thinning of 20 to reduce the correlation might be considered.

4.2.7.1 Monitoring metropolis-like samplers

Samplers that don't necessarily accept a new value at each iteration cannot be monitored as easily as those that do produce new values (such as the Gibbs Sampler). With, for example, a Metropolis-Hastings algorithm the acceptance rate of new proposals is an important measure of the performance of the algorithm. The acceptance rate is defined as the number of iterations where new values are accepted out of a batch of iterations. Let's assume we have a batch size of $n_l = 100$ iterations and during that period we observe m_l accepted proposals. We assume that the number of parameters is small ($p << n_l$) so that there are potentially many transitions that could be made within n_l. The acceptance rate is just $A_r = \frac{m_l}{n_l}$. This rate could be a useful indicator of the behavior of the sampler. For example if the sampler is not mixing well then it may stick in various places failing to find acceptable proposals. This would lead to a low acceptance rate. However, on the other hand, a high acceptance rate may signify good proposals but could also mean that the sampler is "stuck" in the vicinity of a peak in the posterior surface and not searching the space in general. In both cases, the proposals may either be too small or too large to adequately search the space. Usually as a guide to a reasonable acceptance rate, for a M-H algorithm with small dimension (1 to 2 parameters) then $A_r \approx 0.5$ would be reasonable (Robert and Casella, 2005). For higher dimensions ($p > 2$) then $A_r \approx 0.25$ is reasonable. Hence for reversible jump algorithms (which are based on M-H steps with high dimension) then $A_r \approx$ 0.25 might be expected. For Metropolis -Langevin or Langevin -Hastings algorithms (such as used in the R package geoRglm or STAN) that incorporate gradient terms then higher rates are optimal ($A_r \approx 0.6$). It should be borne in mind that in itself achievement of an optimal A_r does not necessarily imply convergence to a stationary distribution, although poor A_r could be due to lack of mixing and hence lack of convergence. It is also possible for chains to have high acceptance and very low convergence (Gamerman and Lopes, 2006). On WinBUGS when a Metropolis update is used then the acceptance rate can be set using the Monitor Met button in the Model Menu. This generates a plot of the acceptance rate over iteration for batches of $n_l = 100$ iterations. For user defined likelihood models using the zeroes or ones trick then A_r is always available.

4.3   Posterior and Likelihood Approximations

From the point of view of computation it is now straightforward to examine a range of posterior distributional forms. This is certainly true for most applications of disease mapping where relative risk is estimated. However there are situations where it may be easier or more convenient to use a form of approximation to the posterior distribution or to the likelihood itself. Some approximations have been derived originally when posterior sampling was not possible and where the only way to obtain fully Bayesian estimates was to approximate (Bernardo and Smith, 1994). However other approximations arise due to the intractability of spatial integrals (for example in point process models).

4.3.1   Pseudo-likelihood and other forms

In certain spatial problems, found in imaging and elsewhere, normalizing constants arise which are highly multidimensional. A simple example is the case of a Markov point process. Define the realization of m events within a window T as $\{s_1, ...s_m\}$. Under a Markov process assumption the normalized probability density of a realization is

$$f_\theta(\mathbf{s}) = \frac{1}{c(\theta)} h_\theta(\mathbf{s})$$

$$\text{where } c(\theta) = \sum_{k=0}^{\infty} \frac{1}{k!} \int_{T^k} h_\theta(\mathbf{s}) \lambda^k (d\mathbf{s}).$$

Conditioning on the number of events (m), then the normalization of $f_m(\mathbf{s}) \propto h_m(\mathbf{s})$ is over the m-dimensional window:

$$c(\theta) = \int_T \ldots \int_T h_m(\{\mathbf{s}_1, \ldots \mathbf{s}_m\}) d\mathbf{s}_1, \ldots d\mathbf{s}_m.$$

For a conditional Strauss process then $f_m(\mathbf{s}) \propto \gamma^{n_R(\mathbf{s})}$ and $n_R(\mathbf{s})$ is the number of R-close pairs of points to $\mathbf{s}$.

It is also true that a range of lattice models developed for image processing applications also have awkward normalization constants (auto-Poisson and autologistic models and Gaussian Markov random field models: Besag and Tantrum, 2003, Rue and Held, 2005).

This has led to the use of approximate likelihood models in many cases. For example, for Markov point processes it is possible to specify a conditional intensity (Papangelou) which is independent of the normalization. This conditional intensity $\lambda^*(\xi, \mathbf{s}|\theta) = h(\boldsymbol{\xi} \cup \mathbf{s})/h(\boldsymbol{\xi})$ can be used within a pseudo-likelihood function. In the case of the above Strauss process this is just $\lambda^*(\xi, \mathbf{s}|\theta) = \lambda^*(\mathbf{s}|\theta) = \gamma^{n_R(\mathbf{s})}$ and the pseudo-likelihood is:

$$L_p(\{\mathbf{s}_1, \ldots \mathbf{s}_m\}|\theta) = \prod_{i=1}^{m} \lambda^*(\mathbf{s}_i|\theta) \exp(-\int_T \lambda^*(\mathbf{u}|\theta) d\mathbf{u}).$$

As this likelihood has the form of an inhomogeneous Poisson process likelihood, then this is relatively straightforward to evaluate. The only issue is the integral of the intensity over the window T. This can be handled via special numerical integration schemes (Berman and Turner, 1992, Lawson, 1992a, Lawson, 1992b, and Section 6.1.1). Bayesian extensions are generally straightforward. Note that once a likelihood contribution can be specified then this can be incorporated within a posterior sampling algorithm such as Metropolis-Hastings. This can be implemented on Win/OpenBUGS via a zeroes trick if the Berman-Turner weighting is used. For example the model with the i th likelihood component: $l_i = \log \lambda^*(\mathbf{s}_i|\theta) - w_i \lambda^*(\mathbf{s}_i|\theta)$ can be fitted using this method, where the weight w_i is based on the Dirichlet tile area of the i th point or a function of the Delauney triangulation around the point (see Berman and Turner, 1992, Baddeley and Turner, 2000, and Appendix C.5.3 of Lawson, 2006).

In application to lattice models, Besag and Tantrum (2003) give the example of a Markov random field of m dimension where the pseudo-likelihood, $L_p = \prod_{i=1}^{m} p(y_i^0|y_{-i}^0; \theta)$ is the product of the full conditional distributions. In the (auto)logistic binary case where the observed binary outcome is y_i^0 then

$$p(y_i^0|y_{-i}^0; \theta) = \frac{\exp(f(\alpha_0, \{y_{-i}^0\}_{\in \partial i}, \theta)}{1 + \exp(f(\alpha_0, \{y_{-i}^0\}_{\in \partial i}, \theta)}$$

where ∂_i denotes the adjacency set of the i th site. This leads to a simple logistic regression conditioned on the sum of the neighboring data values (S_{δ_i}) which should account of at least some of the spatial correlation in the risk distribution.

Other variants of these likelihoods have been proposed. Local likelihood (Tibshirani and Hastie (1987)) is a variant where a contribution to likelihood is defined within a local domain of the parameter space. In spatial problems this could be a spatial area. This has been used in a Bayesian disease mapping setting by Hossain and Lawson (2005). Pairwise likelihood (Nott and Rydén, 1999, Heagerty and Lele, 1998) has been proposed for image restoration and for general spatial mixed models (Varin et al., 2005). Approximate point process likelihoods based on aggregations have been explored by Hossain and Lawson (2008). All these variants of full likelihoods will lead to models that are approximately valid for real applications. It should be born in mind however that they ignore aspects of the spatial correlation and if these are not absorbed in some part of the model hierarchy then this may affect the appropriateness of the model.

4.3.2   Asymptotic approximations

It is possible to approximate a posterior distribution with a simpler distribution which is found asymptotically. The usefulness of approximations lies in their often common form and also the ease with which parameters may be estimated under the approximation. Often the asymptotic approximating distribution will be a normal distribution.

4.3.2.1   Laplace integral approximation

In some situations ratios of integrals must be evaluated and it is possible to employ an integral approximation method suggested by Laplace (Tierney and Kadane, 1986). For example the posterior expectation of a real valued function $g(\boldsymbol{\theta})$ is given by

$$E(g(\boldsymbol{\theta})|\mathbf{y}) = \int g(\boldsymbol{\theta})p(\boldsymbol{\theta}|\mathbf{y})d\boldsymbol{\theta}.$$

This can be considered as a ratio of integrals, given the normalization of the posterior distribution. The approximation is given by

$$\widehat{E}(g(\boldsymbol{\theta})|\mathbf{y}) \approx \left(\frac{\sigma^*}{\sigma}\right) \exp\{-m[h^*(\boldsymbol{\theta}^*) - h(\boldsymbol{\theta})]\}$$

where $-mh(\boldsymbol{\theta}) = \log p(\boldsymbol{\theta}) + l(\mathbf{y}|\boldsymbol{\theta})$ and $-mh^*(\boldsymbol{\theta}) = \log g(\boldsymbol{\theta}) + \log p(\boldsymbol{\theta}) + l(\mathbf{y}|\boldsymbol{\theta})$ and

$$-h(\widehat{\boldsymbol{\theta}}) = \max_{\boldsymbol{\theta}}\{-h(\boldsymbol{\theta})\}, \quad -h^*(\boldsymbol{\theta}^*) = \max_{\boldsymbol{\theta}}\{-h^*(\boldsymbol{\theta})\},$$

$\widehat{\sigma} = |m\nabla^2 h(\widehat{\boldsymbol{\theta}})|^{-1/2}$ and $\widehat{\sigma} = |m\nabla^2 h^*(\boldsymbol{\theta}^*)|^{-1/2}$ where

$$[\nabla^2 h(\widehat{\boldsymbol{\theta}})]_{ij} = \left.\frac{\partial^2 h(\boldsymbol{\theta})}{\partial\theta_i\partial\theta_j}\right|_{\boldsymbol{\theta}=\widehat{\boldsymbol{\theta}}}.$$

This approximation has been adopted by the package R-INLA (chapter 10), and avoids the computational burden of posterior sampling.

5

Bayesian model Goodness of Fit Criteria

As part of any modeling exercise it is usually of interest to assess how well a given model describes given data. To this end a number of measures have been devised to help in this regard. The first of these is a deviance based measure called the deviance information criterion. Second the WAIC or Watanabe Akaike information Criterion and posterior predictive loss and cross validatory measures.

5.1   The Deviance Information Criterion

The *deviance information criterion (DIC)* has been proposed by Spiegelhalter et al. (2002) and is widely used in Bayesian modeling. This is defined as

$$DIC = 2E_{\theta|y}(D) - D[E_{\theta|y}(\theta)],$$

where $D(.)$ is the deviance of the model and y is the observed data. Note that the DIC, for McMC samples, is based on a comparison of the average deviance ($\overline{D} = -2\sum_{g=1}^{G} l(y|\theta^g)/G$) and the deviance of the posterior expected parameter estimates, $\widehat{\theta}$ say: ($\widehat{D}(\widehat{\theta}) = -2l(y|\widehat{\theta})$). For any sample parameter value θ^g the deviance is just $\widehat{D}(\theta^g) = -2l(y|\theta^g)$. The effective number of parameters (pD) is estimated as $\widehat{pD} = \overline{D} - \widehat{D}(\widehat{\theta})$ and then $DIC = \overline{D} + \widehat{pD} = 2\overline{D} - \widehat{D}(\widehat{\theta})$. Unfortunately in some situations the $\widehat{pD}$ can be negative (as it can happen that $\widehat{D}(\widehat{\theta}) > \overline{D}$). Instability in pD can lead to problems in the use of this DIC. For example, mixture models, or more simply, models with multiple modes can "trick" the pD estimate because the overdispersion in such models (when the components are not correctly estimated) leads to $\widehat{D}(\widehat{\theta}) > \overline{D}$ (see e.g., Lunn et al., 2012). For alternative DIC formulations that can allow for mixture models see, e.g., Celeux et al. (2006). It is also true that inappropriate choice of hyper-parameters for variances of parameters in hierarchical models can lead to inflation also, as can nonlinear transformations (such as changing from a Gaussian model to a log normal model). In such cases it is sometimes safer to compute the effective number of parameters from the posterior variance of the deviance. Gelman et al. (2004), p. 182

propose the estimator $\widetilde{pD} = \frac{1}{2}\frac{1}{G-1}\sum_{g=1}^{G}(\widehat{D}(\theta^g) - \overline{D})^2$. This value can also be computed from sample output from a chain. (It is also available directly in R2WinBUGS.) An alternative estimator of the variance is directly available from output: $\widehat{var}(D) = \frac{1}{G-1}\sum_{g=1}^{G}(\widehat{D}(\theta^g) - \overline{D})^2 = 2\widetilde{pD}$. Hence a DIC based on this last variance estimate is just $DIC = \overline{D} + \widehat{var}(D)/2$. Note that the *expected predictive deviance* (EPD: D_{pr}) is an alternative measure of model adequacy and it is based on the out-of-sample predictive ability of the fitted model. The quantity can also be approximately estimated as $\widehat{D}_{pr} = 2\overline{D} - \widehat{D}(\widehat{\theta})$.

5.2   Watanabe AIC (WAIC)

Recently a new measure of goodness of fit has been proposed termed the Watanabe-Akaike Information criterion (WAIC) (see, e.g., Gelman et al., 2014). This can be computed from the log pointwise predictive density (lppd), for G sampled values, as

$$lppd = \sum_{i=1}^{m}\log\{\sum_{g=1}^{G}p(y_i|\theta^g)\}$$

where $p(.|.)$ is the data density. The parameterization penalty is the variance of the log pointwise predictive density summed over the data points to yield the effective number of parameters: $pWAIC = \sum_{i=1}^{m}V_{g=1}^{G}(\log p(y_i|\theta^g))$ where $V_{g=1}^{G}(a) = \frac{1}{G-1}\sum_{g=1}^{G}(a - \overline{a})^2$. This leads to

$$WAIC = lppd - pWAIC.$$

The WAIC can be computed from sampled McMC output and is available automatically within nimble, CARBayes and R-INLA. Spiegelhalter et al. (2014) compare the DIC and WAIC and note the potential instability of DIC and the improvement of the WAIC formulation.

5.3 Posterior Predictive Loss

Gelfand and Ghosh (1998) proposed a loss function based approach to model adequacy which employs the predictive distribution. The approach essentially compares the observed data to predicted data from the fitted model. Define the i th predictive data item as y_i^{pr}. Note that the predictive data can easily be obtained from a converged posterior sample. Given the current parameters at iteration $j : \boldsymbol{\theta}^{(j)}$ say, then

$$p(y_i^{pr}|\mathbf{y}) = \int p(y_i^{pr}|\boldsymbol{\theta}^{(j)})p(\boldsymbol{\theta}^{(j)}|\mathbf{y})d\boldsymbol{\theta}^{(j)}.$$

Hence the j th iteration can yield y_{ij}^{pr} from $p(y_i^{pr}|\boldsymbol{\theta}^{(j)})$. The resulting predictive values has marginal distribution $p(y_i^{pr}|\mathbf{y})$. For a Poisson distribution, this simply requires generation of counts as $y_{ij}^{pr} \leftarrow Pois(e_i\theta_i^{(j)})$.

A loss function is assumed where $L_0(y, y^{pr}) = f(y, y^{pr})$. A convenient choice of loss could be the squared error loss whereby we define the loss as:
$L_0(y, y^{pr}) = (y - y^{pr})^2$.

Alternative loss functions could be proposed such as absolute error loss or more complex (quantile) forms. An overall crude measure of loss across the data is afforded by the average loss across all items: the *mean squared predictive error* (MSPE) is simply an average of the item-wise squared error loss:

$$MSPE_j = \sum_i (y_i - y_{ij}^{pr})^2/m$$

$$and$$
$$MSPE = \sum_i \sum_j (y_i - y_{ij}^{pr})^2/(G \times m),$$

where m is the number of observations and G is the sampler sample size. An alternative could be to specify an absolute error :

$$MAPE_j = \sum_i |y_i - y_{ij}^{pr}|/m$$

$$and$$
$$MAPE = \sum_i \sum_j |y_i - y_{ij}^{pr}|/(G \times m).$$

Gelfand and Ghosh (1998) proposed a more sophisticated form: and m_p is the prediction sample size (usually $G = m_p$). Here, the k can be chosen to weight the different components. For $k = \infty$, then $D_k = A + B$. The choice of k does not usually affect the ordering of model fit. Each component measures a different feature of the fit: A represents lack of fit and B degree of smoothness. The model with lowest D_k (or $MSPE$ or $MAPE$) would be preferred.

Note that predictive data can be easily obtained from model formulae in WinBUGS: in the Poisson case with observed data y[] and predicted data ypred[], we have for the i th item

  y[i]~dpois(mu[i])
  ypred[i]~dpois(mu[i])

As the predicted values are missing they would have to be initialized.

In addition, it is possible to consider prediction-based measures as convergence diagnostics. The measures $\widehat{D}(\theta^{(j)})$, or $MSPE_j$ could be monitored using the single or multi-chain diagnostics discussed above (Section 4.2.6). Note that for any model for which a unit likelihood contribution is available, it is possible to compute a deviance-based measure such as $\widehat{D}(\theta^{(j)})$. Hence for point process (case event), as well as count-based likelihoods, deviance measures are available whereas a residual based measure (such as MSPE) is more difficult to define for a spatial event domain. A further cross-validatory measure of goodness of fit called marginal predictive likelihood (MPL) or log of the MPL (LMPL) is discussed in a later section 5.7.

5.4   Bayesian Residuals

Carlin and Louis (2000) describe a Bayesian residual as:

$$r_i = y_i - \frac{1}{G} \sum_{g=1}^{G} E(y_i|\theta_i^{(g)}) \qquad (5.1)$$

where $E(y_i|\theta_i)$ is the expected value from the posterior predictive distribution, and (in the context of McMC sampling) $\{\theta_i^{(g)}\}$ is a set of parameter values sampled from the posterior distribution.

In the tract count modeling case, with a Poisson likelihood and expectation $e_i\theta_i$, this residual can be approximated, when a constant tract rate is assumed, by:

$$r_i = y_i - \frac{1}{G} \sum_{g=1}^{G} e_i\theta_i^{(g)}. \qquad (5.2)$$

This residual averages over the posterior sample. An alternative computational possibility is to average the $\{\theta_i^{(g)}\}$ sample, $\widehat{\theta}_i = \frac{1}{G}\sum_{g=1}^{G}\theta_i^{(g)}$ say, to yield a posterior expected value of y_i, say $\widehat{y}_i = e_i\widehat{\theta}_i$, and to form $r_i = y_i - \widehat{y}_i$.

A further possibility is to simply form r_i at each iteration of a posterior sampler and to average these over the converged sample (Spiegelhalter et al., 1996. These residuals can provide pointwise goodness-of-fit (*gof*) measures as well as global *gof* measures, (such as mean squared error (MSE): $\frac{1}{m}\sum_{i=1}^{m} r_i^2$) and can be assessed using Monte Carlo methods. For exploratory purposes it might be useful to standardize the residuals before examination, although this is not essential for Monte Carlo assessment. To provide for a Monte Carlo assessment of individual unit residual behavior a repeated Monte Carlo simulation of independent samples from the predictive distribution would be needed. This can be achieved by taking J samples from the converged McMC stream with gaps of length p, where p is large enough to ensure independence. Ranking of the residuals in the pooled set $(J+1)$ can then be used to provide a Monte Carlo p-value for each unit.

5.5 Predictive Residuals and the Bootstrap

It is possible to disaggregate the MSPE to yield individual level residuals based on the predictive distribution. Define $y_i^{pr} \sim f_{pr}(\mathbf{y}^{pr}|\mathbf{y};\boldsymbol{\theta})$ where $f_{pr}(\mathbf{y}^{pr}|\mathbf{y};\boldsymbol{\theta}) = \int f(\mathbf{y}^{pr}|\boldsymbol{\theta})p(\boldsymbol{\theta}|\mathbf{y})d\boldsymbol{\theta}$ and $f(\mathbf{y}^{pr}|\boldsymbol{\theta})$ is the likelihood of $\mathbf{y}^{pr}$ given $\boldsymbol{\theta}$. This can be approximated within a converged sample by a draw from $f(\mathbf{y}^{pr}|\boldsymbol{\theta})$. For a Poisson likelihood, at the g th iteration, with expectation $e_i\theta_i^{(g)}$, a single value $y_i^{pr(g)}$ is generated from $Pois(e_i\theta_i^{(g)})$. Hence a predictive residual can be formed from $r_i^{pr} = y_i - y_i^{pr}$. This must be averaged over the sample. This can be done in a variety of ways. For example, we could take

$$r_i^{pr} = \sum_{g=1}^{G} \{y_i - y_i^{pr(g)}\}/G.$$

Other possibilities could be explored (see for example, Marshall and Spiegelhalter, 2003).

To further assess the distribution of residuals, it would be advantageous to be able to apply the equivalent of PB in the Bayesian setting. With convergence of a McMC sampler, it is possible to make subsamples of the converged output. If these samples are separated by a distance (h) which will guarantee approximate independence (Robert and Casella, 2005), then a set of J such samples could be used to generate $\{y_i^{pr}\}$ $j = 1, ..., J$, with, $y_i^{pr} \leftarrow Pois(e_i\widehat{\theta}_{ij})$, and the residual computed from the data r_i can be compared to the set of J residuals computed from $y_i^{pr} - \widehat{y}_i$. In turn, these residuals can be used to assess functions of the residuals and gof measures. The choice of J will usually be 99 or 999 depending on the level of accuracy required.

5.6   Conditional Predictive Ordinates (CPO)

It is possible to consider a different approach to inference whereby individual observations are compared to the predictive distribution with observations removed. This conditional approach has a cross-validation flavor, i.e., the value in the unit is predicted from the remaining data and compared to the observed data in the unit. The derived residual is defined as

$$r_i^{CPO} = y_i - y_{i,-i}^{rep}$$

where $y_{i,-i}^{rep}$ is the predicted value of y based on the data with the i th unit removed (Stern and Cressie (2000)). The value of $y_{i,-i}^{rep}$ is obtained from the cross-validated posterior predictive distribution:

$$p(y_{i,-i}^{rep}|\mathbf{Y}_{-i}) = \int p(y_{i,-i}^{rep}|\boldsymbol{\theta})p(\boldsymbol{\theta}|\mathbf{Y}_{-i})d\boldsymbol{\theta}$$

where $\boldsymbol{\theta}$ is a vector of model parameters. For a Poisson data likelihood, $p(y_{i,-i}^{rep}|\mathbf{Y}_{-i})$ is just a Poisson distribution with mean $e_i^*\theta_i$ where e_i^* is adjusted for the removal of the i th unit and θ_i is estimated under the cross-validated posterior distribution $p(\boldsymbol{\theta}|\mathbf{Y}_{-i})$.

As noted by Spiegelhalter et al. (1996), it is possible to make inference about such residuals within a conventional McMC sampler via the construction of weights. Assume draws $g = 1, ..., G$ are available and define the importance weight $w_{-i}(\boldsymbol{\theta}^g) = \frac{1}{p(y_i|e_i\theta_i^g)}$. This is just the reciprocal of the Poisson probability with mean $e_i\theta_i^g$. It is then possible to compute a Monte Carlo probability for the residual via:

$$p(y_{i,-i}^{rep} \leq y_i|\mathbf{Y}_{-i}) = p(y_{i,-i}^{rep} < y_i|\mathbf{Y}_{-i}) + \frac{1}{2}p(y_{i,-i}^{rep} = y_i|\mathbf{Y}_{-i})$$

$$\approx \sum_{l=0}^{y_i-1}\left\{\sum_{g=1}^{G}p(y_{i,-i}^{rep} = g|\boldsymbol{\theta}^g)w_{-i}(\boldsymbol{\theta}^g)\right\}/w_i^T$$

$$+\frac{1}{2}\left\{\sum_{g=1}^{G}p(y_{i,-i}^{rep} = y_i|\boldsymbol{\theta}^g)w_{-i}(\boldsymbol{\theta}^g)\right\}/w_i^T$$

where $w_i^T = \sum_{g=1}^{G}\frac{1}{p(y_i|e_i\theta_i^g)}$. In general, a simple approach to computation of the CPO without recourse to refitting is to note that the conditional predictive ordinate for the i the unit can be obtained from

$$CPO_i^{-1} = G^{-1}\sum_{g=1}^{G}p(y_i|\boldsymbol{\theta}^g)^{-1}$$

where $p(y_i|\boldsymbol{\theta}^g)$ is the data density given the current parameters. Note that the CPO_i can be used as local measure of model fit in that $0 < CPO_i < 1$

and values closest to 1 suggest good fit while values close to zero suggest data points that are poorly fitted. An early example of the use of these diagnostics is found in Dey et al. (1997).

5.7 Pseudo Bayes Factors and Marginal Predictive Likelihood

It has been observed that it is possible to compute an approximation to a Bayes Factor based on CPO_i values (Gelfand and Dey, 1994; Dey et al., 1997). For two different models $M1$ and $M2$ the pseudo Bayes factor for $M1$ versus $M2$ can be written as $PBF = \prod_i CPO_i(M1)/\prod_i CPO_i(M2)$, and for a given model ($M1$ say) then $PL(M1) = \sum_i \log(CPO_i(M1))$ is an estimate of the log marginal predictive likelihood (LMPL) for that model. A comparison can be made between any competing models based on values of $PL(.)$. In general, the least negative the value of $PL(.)$ the higher the probability of the model. LMPL, DIC, WAIC, and MSPE are all commonly used as overall model goodness of fit measures. LMPL is available on INLA and CARBayes, while the DIC is available on Win/OpenBUGS and INLA, and the WAIC on CARBayes, INLA, and nimble.

5.8 Exceedance Probabilities

Exceedance probabilities are important when assessing the localized spatial behavior of the model and the assessment of unusual clustering or aggregation of disease. The simplest case of an exceedance probability is $q_i^c = \Pr(\theta_i > c)$. The probability is an estimate of how frequently the relative risk exceeds the null risk value ($\theta_i = 1$) and can be regarded as an indicator of "how unusual" the risk is in that unit. This leads to assessment of "hot spot" clusters: areas of elevated risk found independently of any cluster grouping criteria. Note that under posterior sampling, a converged sample of $\{\theta^{m+1}, ..., \theta^{m+m_p}\}$ can yield posterior expected estimates of these probabilities as $\hat{q}_i^c = \sum_{g=m+1}^{m+m_p} I(\theta_i^{(g)} > c)/G$ where $G = m_p$. It is straightforward on Win/OpenBUGS or nimble, for example, to compute these values. If theta[i] is set to store the current θ_i then

```
prexc[i]<-step(theta[i]-1)
```

will store the indicator of exceedance and this variable can be monitored. The posterior average of the variable will yield the posterior estimate of the probability. Large values of $\hat{q}_i^c$ would be suggestive of unusual areas of risk. In fact this measure has been proposed as a method for detecting clusters (Richardson et al., 2004; Abellan et al., 2008; Hossain and Lawson, 2006, 2010). While it certainly can be used to examine maps for individual hot spots of risk, the measure itself does not directly measure clustering of risk in terms of spatial aggregation. In fact, the measure can be applied to any underlying model when a relative risk parameter is estimated and hence may be model dependent.

There are a number of issues associated with the evaluation of $\hat{q}_i^c$. First the value of c must be specified. Second the level of probability that is regarded *"unusual"* (a say) must also be fixed. However there is a trade-off between these measures. Not only will there be different features found if you change the threshold (c), but there will be different interpretations if threshold is changed from say $a_1 = 0.95$ to $a_2 = 0.99$ or lowered to $a_0 = 0.90$. Hence a simple rule such as, *classify as unusual any region where* $\hat{q}_i^1 > a_1$, might be equivalent to $\hat{q}_i^2 > a_0$. Whereas the null level of risk may be $c = 1$ there is no reason to assume as a threshold $\hat{q}_i^1 > 0.95$. Of course usually either a or c is fixed. It should be noted that $\hat{q}_i^c$ is a function of a model and so is not necessarily going to yield the same information as, say a residual r_i. While both depend on model elements, a residual usually also contains extra (at least) uncorrelated noise and should, if the model fits well, not contain any further structure. On the other hand, posterior estimates of relative risk will include modelled components of risk (such as trend or correlation) and should be relatively free of extra noise.

Richardson et al. (2004) and later Abellan et al. (2008) describe the use of exceedance within environmental health studies. Exceedance is now use commonly to identify hot spot clustering of disease risk, and more widely in other application areas to identify extreme events. It has been demonstrated to be superior in identification of clusters compared to commonly used scanning methods (Zou et al., 2018). Neighborhood exceedances have also been examined (Hossain and Lawson, 2006, Hossain and Lawson, 2010).

6

Bayesian Disease Mapping Models

In this chapter I will review some basic disease mapping models and their implementation.

6.1 An Introduction to Case Event and Count Likelihoods

6.1.1 The Poisson process model

Define a study area as T and within that area m event of disease occur. These events are usually address locations of the cases. The case could be an incident or prevalent case or could be a death certificate address. We assume at present that the cases are geo-coded down to a point (with respect to the scale of the total study region). Hence they form a point process in space. Define $\{s_i\}, i = 1, ..., m$ as the set of all cases within T. This is called a *realization* of the disease process, in that we assume that all cases within the study area are recorded. This is a common form of data available from government agencies.

The basic point process model assumed for such data within disease mapping is the heterogeneous Poisson process with first order intensity $\lambda(s)$. The basic assumptions of this model are that points (case events) are independently spatially-distributed and governed by the first order intensity. Due to the independence assumption, we can derive a likelihood for a realization of a set of events within a spatial region. For the study region defined above the unconditional likelihood of m events is just

$$L(\{\mathbf{s}\}|\psi) = \frac{1}{m!} \prod_{i=1}^{m} \lambda(s_i|\psi) \exp\{-\Lambda_T\} \tag{6.1}$$

$$\text{where } \Lambda_T = \int_T \lambda(u|\psi)du.$$

The function Λ_T is the integral of the intensity over the study region, ψ is a parameter vector, and $\lambda(s_i|\psi)$ is the first order intensity evaluated at the case event location s_i. Denote this likelihood as $PP[\{\mathbf{s}\}|\psi]$. This likelihood can be maximized with respect to the parameters in ψ and likelihood-based inference could be pursued. The only difficulty in its evaluation is the estimation of the spatial integral. However a variety of approaches can be used for numerical integration of this function and with suitable weighting schemes this likelihood can be evaluated even with conventional linear modeling functions within software packages (such as glm in R) (see e.g., Berman and Turner, 1992, Lawson, 2006, App. C). An example of such a weighted log-likelihood approximation is:

$$l(\{\mathbf{s}\}|\psi) = \sum_{i=1}^{m} \ln \lambda(s_i|\psi) - \Lambda_T, \tag{6.2}$$

where $\Lambda_T \approx \sum_{i=1}^{m} w_i \lambda(s_i|\psi)$ and w_i is an integration weight. This scheme per se is not accurate and more weights are needed. In the more general scheme of Berman & Turner a set of additional mesh points (of size m_{aug}) are added to the data. The augmented set $(N = m + m_{aug})$ is used in the likelihood with an indicator function, I_k:

$$l(\{\mathbf{s}\}|\psi) = \sum_{k=1}^{N} w_k \{\frac{I_k}{w_k} \ln \lambda(s_k|\psi) - \lambda(s_k|\psi)\},$$

$$\text{where} \int_T \lambda(u|\psi)du = \sum_{k=1}^{N} w_k \lambda(s_k|\psi).$$

This has the form of a weighted Poisson likelihood, with $I_k = 1$ for a case and 0 otherwise. Diggle (1990) gives an example of the use of a likelihood such as (6.1) in a spatial health data problem.

In disease mapping applications, it is usual to parameterize $\lambda(s|\psi)$ as a function of two components. The first component makes allowance for the underlying population in the study region, and the second component is usually specified with the modelled components (i.e., those components describing the "excess" risk within the study area).

A typical specification would be

$$\lambda(s|\psi) = \lambda_0(s|\psi_0).\lambda_1(s|\psi_1). \tag{6.3}$$

Here it is assumed that $\lambda_0(s|\psi_0)$ is a spatially-varying function of the population "at risk" of the disease in question. It is parameterized by ψ_0. The second function, $\lambda_1(s|\psi_1)$, is parameterized by ψ_1 and includes any linear or non-linear predictors involving covariates or other descriptive modeling terms thought appropriate in the application. Often we assume, for positivity, that $\lambda_1(s_i|\psi_1) = \exp\{\eta_i\}$ where η_i is a parameterized linear predictor allowing a link to covariates measured at the individual level. The covariates could include spatially-referenced functions as well as case-specific measures. Note that $\psi : \{\psi_0, \psi_1\}$. The function $\lambda_0(s|\psi_0)$ is a nuisance function which must be allowed for but which is not usually of interest from a modeling perspective.

6.1.2 The conditional logistic model

When a bivariate realization of cases and controls are available it is possible to make conditional inference on this joint realization. This leads to a simpler conditional logistic model formulation. Define the case events as $s_i : i = 1, ..., m$ and the control events as $s_i : i = m + 1, ..., N$ where $N = m + n$ the total number of events. Associated with each location is a binary variable (y_i) which labels the event either as a case ($y_i = 1$) or a control ($y_i = 0$). Assume also that the point process models governing each event type (case or control) is a heterogeneous Poisson process with intensity $\lambda(s|\psi)$ for cases and $\lambda_0(s|\psi_0)$ for controls. The superposition of the two processes is also a heterogeneous Poisson process with intensity $\lambda_0(s|\psi_0) + \lambda(s|\psi) = \lambda_0(s|\psi_0)[1 + \lambda_1(s|\psi_1)]$. Conditioning on the joint realization of these processes, then it is straightforward to derive the conditional probability of a case at any location as

$$\Pr(y_i = 1) = \frac{\lambda_0(s_i|\psi_0).\lambda_1(s_i|\psi_1)}{\lambda_0(s_i|\psi_0)[1 + \lambda_1(s_i|\psi_1)]}$$

$$= \frac{\lambda_1(s_i|\psi_1)}{1 + \lambda_1(s_i|\psi_1)} = p_i \qquad (6.4)$$

and

$$\Pr(y_i = 0) = \frac{1}{1 + \lambda_1(s_i|\psi_1)} = 1 - p_i. \qquad (6.5)$$

The important implication of this result is that the background nuisance function $\lambda_0(s_i|\psi_0)$ drops out of the formulation and, further, this formulation leads to a standard logistic regression if a linear predictor is assumed. For example, a log linear formulation for $\lambda_1(s_i|\psi_1)$ leads to a logit link to p_i, i.e.,

$$p_i = \frac{\exp(\eta_i)}{1 + \exp(\eta_i)},$$

where $\eta_i = x_i'\beta$ and x_i' is the i th row of the design matrix of covariates and β is the corresponding p-length parameter vector. Note that slightly different formulations can lead to non-standard forms (see e.g., Diggle and Rowlingson, 1994). In some applications, non-linear links to certain covariates may be appropriate. Further, if the probability model in (6.4) applies, then the likelihood of the realization of cases and controls is simply

$$L(\psi_1|\mathbf{s}) = \prod_{i \in cases} p_i \prod_{i \in controls} 1 - p_i$$

$$= \prod_{i=1}^{N} \left[\frac{\{\exp(\eta_i)\}^{y_i}}{1 + \exp(\eta_i)} \right]. \qquad (6.6)$$

Hence, in this case, the analysis reduces to that of a logistic likelihood, and this has the advantage that the "at risk" population nuisance function does not have to be estimated. This model is ideally suited to situations where it is natural to have a control and case realization, where conditioning on the spatial pattern is reasonable. A fuller review of case event models in this context is found in Lawson (2012) and Onicescu and Lawson (2016). Examples where this type of analysis could be important are often found in environmental studies where locations of cases and controls are thought to reflect the influence of adverse environmental conditions, and residential address or some similarly fine resolution geocode (e.g., census block) is available for the observed data. An example of this modeling with respiratory cancers is found in Chapter 17.

6.1.3   The binomial model for count data

In the case where we examine arbitrary small areas (such as census tracts, counties, postal zones, municipalities, health districts), usually a count of disease is observed within each spatial unit. Define this count as y_i and assume that there are m small areas. We also consider that there is a finite population within each small area out of which the count of disease has arisen. Denote this as $n_i \ \forall_i$. In this situation, we can consider a binomial model for the count data conditional on the observed population in the areas. Hence we can assume that given the probability of a case is p_i, then y_i is distributed independently as

$$y_i \sim bin(p_i, n_i)$$

and that the likelihood is given by

$$L(y_i|p_i, n_i) = \prod_{i=1}^{m} \binom{n_i}{y_i} p_i^{y_i} (1 - p_i)^{(n_i - y_i)}. \tag{6.7}$$

It is usual for a suitable link function for the probability p_i to a linear predictor, η_i, to be chosen. The commonest would be a logit link so that

$$p_i = \frac{\exp(\eta_i)}{1 + \exp(\eta_i)}.$$

Here, we envisage the model specification within η_i, to include spatial and non-spatial components. Two applications which are well suited to this approach are the analysis of sex ratios of births, and the analysis of birth outcomes (e.g., birth abnormalities) compared to total births. Sex ratios are often derived from the number of female (or male) births compared to the total birth population count in an area. A ratio is often formed, though this is not necessary in our modeling context. In this case the p_i will often be close to 0.5 and spatially-localized deviations in p_i may suggest adverse environmental risk (Williams et al., 1992). The count of abnormal births (these could include any abnormality found at birth) can be related to total births in an area. Variations in abnormal birth count could relate to environmental as well as health service variability (over time and space)(Morgan et al., 2004).

6.1.3.1 Bernoulli model

A special case of the binomial model arises when the disease outcome (y_i) is binary and so the binomial model simplifies to a Bernoulli model, i.e., $y_i \sim Bern(p_i)$. In this case individual level predictors could be available (such as age, gender, ethnicity, socio-economic status(SES)) but the geocoding could be contextual in that it may be the case that only an aggregated spatial unit can be used. For example, in the US it is common for the individual patients to be associated with a zipcode (postal district code), but no exact address or finer scale geocoding. It is very common in observational studies to have sampled individuals with their associated individual covariates/predictors and only a large scale geocoding, such as zipcode, or census tract, within county or province within state. In this way a contextual hierarchy of (neighborhood) effects is available, but often in a fragmented geography due to the sampling frame used. Hence the model for p_i could be

$$
\text{logit}(p_i) = f_i(\text{individual predictors}) + g_i(\text{zip, tract, county covariates.....})
$$
$$
+ R_i(\text{zip, tract, county...........}) \tag{6.8}
$$

where g_i, R_i denotes respectively the predictor effect and random effect pertaining to the individual residing in that spatial unit.

6.1.4 The Poisson model for count data

Perhaps the most commonly encountered model for small area count data is the Poisson model. This model is appropriate when there is a relatively low count of disease and the population is relatively large in each small area. Often the disease count y_i is assumed to have a mean μ_i and is independently distributed as

$$
y_i \sim Poisson(\mu_i). \tag{6.9}
$$

The likelihood is given by

$$
L(\mathbf{y}|\boldsymbol{\mu}) = \prod_{i=1}^{m} \mu_i^{y_i} \exp(-\mu_i)/y_i!. \tag{6.10}
$$

The mean function is usually considered to consist of two components: (i) a component representing the background population effect, and (ii) a component representing the excess risk within an area. This second component is often termed the *relative risk*. The first component is commonly estimated or computed by comparison to rates of the disease in a standard population and a local expected rate is obtained. This is often termed *standardization* (Inskip et al., 1983). Hence, we would usually assume that the data is independently distributed with expectation

$$
E(y_i) = \mu_i = e_i \theta_i
$$

where e_i is the expected rate for the i th area and θ_i is the relative risk for the i th area. As we will be developing Bayesian hierarchical models we will consider $\{y_i\}$ to be conditionally independent given knowledge of $\{\theta_i\}$. The expected rate is usually assumed to be fixed for the time period considered in the spatial example, although there is literature on the estimation of small area rates that suggests this may be naive (Ghosh and Rao, 1994, Rao, 2003).

6.1.4.1   Standardization

The expected rates or counts that are used in Poisson models are usually assumed to be fixed quantities. They are a product of *indirect* standardization. (The alternative to indirect standardization is direct standardization which converts the outcome counts to rates thereby changing the data model.) For indirect standardization, which is the commonest form of adjustment, a standard population is chosen to which the disease outcome is to be compared. In the standard population the overall disease rate is computed and then applied to each local area based on their population size. For example,

$$R = \frac{\sum y_i}{\sum p_i} \text{ and } e_i = p_i R$$

where p_i is the local area population and the sum is over all relevant areas. Essentially this rate proportionately allocates the disease risk based on local population. The choice of standard population, over which to calculate the rate R, is crucial for the computation of the expected rate/count. Often for spatial studies the whole study area is used as reference, as the internal relative changes in risk are of importance. However, larger areas could also be used or indeed sometimes an area external to the study region could be chosen. For example, a study of risk within New Jersey could have as standard population the whole of the north east US. An external reference population could be useful when two study areas are to be compared. For example, if a comparison of New Jersey to Pennsylvania is to be made then New England could be the reference, say. Alternatively the combined states could act as reference for each. The choice of reference standard population is crucial and different choices could lead to different conclusions concerning risk. In most of studies the expected rate/count is assumed to be fixed once computed. Note also that more sophisticated standardization can be used whereby sub strata in the population can be adjusted for. For instance age and gender distribution can be included as part of the adjustment of the expected rate. This could be important when there is a strong spatial differentiation in sub strata distribution across a study area.

6.1.4.2   Relative risk

Usually the focus of interest will be the modeling of the relative risk. The commonest approach to this is to assume a logarithmic link to a linear predictor

model:
$$\log \theta_i = \eta_i.$$
This form of model has seen widespread use in the analysis of small area count data in a range of applications (see e.g., Leyland and Goldstein, 2001, ch. 10, Stevenson et al., 2005, Waller and Gotway, 2004, ch. 10).

6.2 Specification of the Predictor in Case Event and Count Models

In all the above models a predictor function (η_i) was specified to relate to the mean of the random outcome variable, via a suitable link function. Often the predictor function is assumed to be linear and a function of fixed covariates and also possibly random effects. We define this in a general form, for p covariates, here as

$$\eta_i = x_i'\boldsymbol{\beta} + z_i'\boldsymbol{\xi},$$
$$= \beta_1 x_{1i}........\beta_p x_{pi} + \xi_1 z_{1i}......\xi_q z_{qi}$$

where x_i' is the i th row of a covariate design matrix $\mathbf{x}$ of dimension $m \times p$, $\boldsymbol{\beta}$ is a $(p \times 1)$ vector of regression parameters, $\boldsymbol{\xi}$ is a $(q \times 1)$ unit vector and z_i' is a row vector of individual level random effects, of which there are q. In this formulation the unknown parameters are $\boldsymbol{\beta}$ and $\mathbf{z}$ the $(m \times q)$ matrix of random effects for each unit. Note that in any given application it is possible to specify subsets of these covariates or random effects. Covariates for case event data could include different types of specific level measures such as an individual's age, gender, smoking status, health provider, etc., or could be environmental covariates which may have been interpolated to the address location of the individual (such as soil chemical measures or air pollution levels) or to a contextual geocoding unit such as zip code etc. For count data in small areas, it is likely that covariates will be obtained at the small area level. For example, for census tracts, there is likely to be socioeconomic variables such as poverty (% population below an income level), car ownership, median income level, available from the census. In addition, some variates could be included as supra-area variables such as health district in which the tract lies. Environmental covariates could also be interpolated to be used at the census tract level. For example air pollution measures could be averaged over the tract.

In some special applications non-linear link functions are used, and in others, mixtures of link functions are used. One special application area where this is found is the analysis of putative hazards (see Chapter 17), where specific distance- and/or direction-based covariates are used to assess evidence for a relation between disease risk and a fixed (putative) source of health hazard. For example, one simple example of this is the conditional logistic modeling of disease cases around a fixed source. Let distance and direction from the source to the i th location be d_i and ϕ_i respectively, then a mixed linear and non-linear link model is commonly assumed where

$$\eta_i = \{1 + \beta_1 \exp(-\beta_2 d_i)\}.\exp\{\beta_0 + \beta_3 \cos(\phi_i) + \beta_4 \sin(\phi_i)\}.$$

Here the distance effect link is nonlinear, while the overall rate (β_0) and directional components are log-linear. The explanation and justification for this formulation is deferred to Chapter 17.

Fixed covariate models can be used to make simple descriptions of disease variation. In particular it is possible to use the spatial coordinates of case events (or in the case of count data, centroids of small areas) as covariates. These can be used to model the long range variation of risk: *spatial trend.* For example, let's assume that the i th unit x-y coordinates are (x_{si}, y_{si}). We could define a polynomial trend model such as:

$$\eta_i = \beta_0 + \sum_{l=1}^{L} \beta_{xl}x_{si}^l + \sum_{l=1}^{L} \beta_{yl}y_{si}^l + \sum_{k=1}^{K}\sum_{l=1}^{L} \beta_{lk}x_{si}^l y_{si}^k.$$

This form of model can describe a range of smoothly varying non-linear surface forms. However, except for very simple models, these forms are not parsimonious and also cannot capture the extra random variation that often exists in disease incidence data.

6.2.1 Regression model formulations

In the Bayesian paradigm all parameters are stochastic and are therefore assumed to have prior distributions. Hence in the regression model

$$\eta_i = x_i'\beta,$$

the β parameters are assumed to have prior distributions. Hence this can be formulated as

$$P(\beta, \tau_\beta | data) \propto L(data|\beta, \tau_\beta) f(\beta | \tau_\beta)$$

where $f(\beta|\tau_\beta)$ is the joint distribution of the covariate parameters conditional on the hyperparameter vector τ_β. Often we regard these parameters as independent and so

$$f(\beta|\tau_\beta) = \prod_{j=1}^{p} f_j(\beta_j | \tau_{\beta_j}).$$

More generally it is commonly assumed that the covariate parameters can be described by a Gaussian distribution and if the parameters are allowed to be correlated then we could have the multivariate Gaussian specification:

$$f(\beta|\tau_\beta) = N_p(0, \Sigma_\beta),$$

where under this prior assumption, $E(\boldsymbol{\beta}|\boldsymbol{\tau}_\beta) = \mathbf{0}$ and $\boldsymbol{\Sigma}_\beta$ is the conditional covariance of the parameters. The commonest specification assumes prior independence and is:

$$f(\boldsymbol{\beta}|\boldsymbol{\tau}_\beta) = \prod_{j=1}^{p} N(0, \tau_{\beta_j}),$$

where $N(0, \tau_{\beta_j})$ is a zero mean single variable Gaussian distribution with variance τ_{β_j}. At this point an assumption about variation in the hyperparameters is usually made. At the next level of the hierarchy hyperprior distributions are usually assumed for $\boldsymbol{\tau}_\beta$. The definition of these distributions could be important in defining the model behavior. For example if a vague hyperprior is assumed for τ_{β_j} this may lead to extra variation being present when limited learning is available from the data. This can affect computation of goodness of fit (GOF) measures and convergence diagnostics. While uniform hyperpriors (on a large positive range) can lead to improper posterior distributions, it has been found that a uniform distribution for the standard deviation can be useful (Gelman, 2006), i.e., $\sqrt{\tau_{\beta_j}} \sim U(0, A)$ where A has a large positive value. Alternative suggestions are usually in the form of gamma or inverse gamma distributions with large variances. For example, Kelsall and Wakefield (2002) proposed the use of gamma(0.2, 0.0001) with expectation 2000 and variance 20,000,000, whereas Banerjee et al. (2004) examine various alternative specifications including gamma (0.001, 0.001). One common specification (Thomas et al., 2004) is gamma (0.5, 0.0005) which has expectation 1000 and variance 2,000,000. A default specification for a precision of a Gaussian distribution on R-INLA is gamma(1, 0.00005) which also has a large variance. Weakly informative priors have also been suggested, such as gamma(2, 1) or gamma($2, 1/A$) where A is the reasonable span of the parameter (Simpson et al., 2015).

While these prior specifications lead to relative uninformativeness, their use has been criticized by Gelman (2006) (see also Lambert et al., 2005), in favor of half-Cauchy and uniform prior distributions on the standard deviation. See also https://github.com/stan-dev/stan/wiki/Prior-Choice-Recommendations for other discussion.

To summarize the hierarchy for such covariate models, often a graphical display of the conditional model structure is employed. This is usually known as a directed acyclic graph (DAG). It is possible to produce these displays on Win/OpenBUGS or R (dagify, ggdag). Figure 6.1 displays a directed acyclic graph for a simple Bayesian hierarchical covariate model with two covariates (x_1, x_2) and relative risk defined as $\theta_i = \exp(\beta_0 + \beta_1 x_{1i} + \beta_2 x_{2i})$ for data $\{y_i, e_i\}$ for m regions. The regression parameters are assumed to have independent zero mean Gaussian prior distributions with fixed precisions.

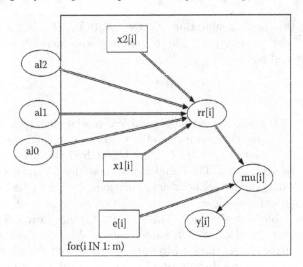

FIGURE 6.1
A directed acyclic graph (in Win/OpenBUGS Doodle format) for a simple Poisson Bayesian regression with log linear relative risk and two covariates.

6.3   Simple Case and Count Data Models with Uncorrelated Random Effects

In the previous section, some simple models were developed. These consisted of functions of fixed observed covariates. In a Bayesian model formulation all parameters are stochastic and so the extension to the addition of random effects is relatively straightforward. In fact the term "mixed" model (linear mixed model: LMM, normal linear mixed model: NLMM, or generalized linear mixed model: GLMM) is strictly inappropriate as there are no fixed effects in a Bayesian model.

The simple regression models described above often do not capture the extent of variation present in count data. Overdispersion or spatial correlation due to unobserved confounders will usually not be captured by simple covariate models and often it is appropriate to include some additional term or terms in a model which can capture such effects.

Initially, overdispersion or extra-variation can be accommodated by either (a) inclusion of a prior distribution for the relative risk, (such as a gamma-Poisson model) or (b) by extension of the linear or non-linear predictor term to include an extra random effect (log-normal model). In both cases the model addresses uncorrelated heterogeneity (UH) in the outcome.

6.3.1 Gamma and beta models

6.3.1.1 Gamma-Poisson models

The simplest extension to the likelihood model that accommodates extra variation is one in which the parameter of interest in the likelihood is given a prior distribution. One case event example would be where the intensity is specified at the i th location as, $\lambda_0(s_i|\psi_0).\lambda_1(s|\psi_1) \equiv \lambda_{0i}.\lambda_{1i}$, suppressing the parameter dependence, for simplicity. Note that λ_{1i} plays the role of a relative risk parameter. This parameter can be assigned a prior distribution, such as a gamma distribution to model extra-variation. For most applications where count data are commonly found a Poisson likelihood is assumed. We will focus on these models in the remainder of this section. The Poisson parameter θ_i could be assigned a $gamma(a, b)$ prior distribution. In this case, the prior expectation and variance would be respectively a/b and a/b^2. This could allow for extra variation or overdispersion. With a Poisson data model and gamma prior distribution for the risk then a gamma-Poisson model results. For fixed a, and b the posterior distribution is known and can be computed directly as a gamma distribution with parameters $y_i + a$, and $e_i + b$. Samples from this posterior distribution are straightforwardly obtained (e.g., using the rgamma function on R). The prior predictive distribution of $\mathbf{y}^*$ is also relevant in this case as it leads to a distribution often used for overdispersed count data: the negative binomial.

6.3.1.1.1 Hyperprior distributions One extension to the above model is to consider a set of hyperprior distributions for the parameters of the gamma prior (a and b). Often these are assumed to also have prior distributions on the positive real line such as gamma (a', b') with $a' > 0$, $b' = 1$ or $b' > 1$. Note that only if hyperprior distributions are assumed for a and b can they be estimated using the posterior distribution.

6.3.1.1.2 Linear parameterization One approach to incorporating more sophisticated model components into the relative risk model is to model the parameters of the gamma prior distribution. For example, gamma linear models can be specified where

$$[\theta_i|y_i, e_i, a, b_i] \sim gamma(a, b_i)$$
$$\text{where } b_i = a/\mu_i \text{ and}$$
$$\mu_i = \eta_i.$$

In this formulation the prior expectation is μ_i and the prior variance is μ_i^2/a. While this formulation could be used for modeling, often the direct linkage between the variance and mean is a disadvantage. As will be seen later, a log normal parameterization is often more useful.

6.3.1.2   Beta-binomial models

When Bernoulli or binomial likelihood models are assumed (such as (6.6) or (6.7)) then one may need to consider prior distributions for the probability parameter p_i. Commonly a beta prior distribution is assumed for this :

$$[p_i | \alpha, \beta] \sim beta(\alpha, \beta).$$

This distribution can flexibly specify a range of forms of distribution from peaked ($\alpha = \beta, \beta > 1$) to uniform ($\alpha = \beta = 1$) and U-shaped ($\alpha = \beta = 0.5$) to skewed or either monotonically decreasing or increasing. In the case of the binomial distribution this prior distribution, with α, β fixed, leads to a posterior distribution of the form:.

$$[\mathbf{p} | \mathbf{y}, \mathbf{n}, \alpha, \beta] = B(\alpha, \beta)^{-m} \prod_{i=1}^{m} [\binom{n_i}{y_i} p_i^{y_i + \alpha - 1} (1 - p_{i'})^{(n_i - y_i + \beta - 1)}].$$

6.3.1.2.1   Hyperprior distributions
The parameters α and β are strictly positive and these could also have hyperprior distributions. However, unless these parameters are restricted to the unit interval, then distributions such as the gamma, exponential or inverse gamma or inverse exponential would have to be assumed as hyperprior distributions.

6.3.1.2.2   Linear parameterization
An alternative specification for modeling covariate effects is to specify a linear or non-linear predictor with a link to a parameter or parameters. For example, it is possible to consider a parameterization such as $\alpha_i = \exp(\eta_i)$ and $\beta_i = \psi \alpha_i$ where ψ is a linkage parameter with prior mean given by $\frac{1}{1+\psi}$. When $\psi = 1$, then the distribution is symmetric. The disadvantage with this formulation is that a single parameter is assigned to the linear predictor and a dependence is specified between α_i and β_i. One possible alternative is to model the prior mean as $\text{logit}(\frac{\alpha_i}{\alpha_i + \beta_i}) = \eta_i$. However this also forces a dependence between α_i and β_i.

6.3.2   Log-normal/logistic-normal models

One simple device that is very popular in disease mapping applications is to assume a direct linkage between a linear or non-linear predictor (η_i) and the parameter of interest (such as θ_i or p_i). This offers a convenient method of introducing a range of covariate effects and unobserved random effects within a simple formulation. The general structure of this formulation is $\eta_i = x_i' \beta + z_i' \gamma$. The simplest form involving uncorrelated heterogeneity would be

$$\eta_i = \beta_0 + z_{1i}$$

where z_{1i} is an uncorrelated random effect. This essentially represents an overall level (β_0) and a random noise term allowing the regions to vary randomly around the overall level.

6.4 Correlated Heterogeneity Models

Uncorrelated heterogeneity (UH) models with gamma or beta prior distributions for the relative risk are useful but have a number of drawbacks. First, as noted above, a gamma distribution does not easily provide for extensions into covariate adjustment or modeling, and, second, there is no simple and adaptable generalization of the gamma distribution with spatially correlated parameters. Wolpert and Ickstadt (1998) provided an example of using correlated gamma field models, but these models have been shown to have poor performance under simulated evaluation (Best et al., 2005). The advantages of incorporating a Gaussian specification are many. First, a random effect which is log-Gaussian behaves in a similar way to a gamma variate, but the Gaussian model can include a correlation structure. Hence, for the case where it is suspected that random effects are correlated, then it is simpler to specify a log Gaussian form for *any* extra variation present. The simplest extension is to consider additive components describing different aspects of the variation thought to exist in the data.

For a spatial Gaussian process, (Ripley, 1981, pg. 10), any finite realization has a multivariate normal distribution with mean and covariance inherited from the process itself, i.e., $\mathbf{x} \sim MVN(\boldsymbol{\mu}, K)$, where $\boldsymbol{\mu}$ is an m length mean vector and K is an $m \times m$ positive definite covariance matrix.

There are many ways of incorporating such heterogeneity in models, and some of these are reviewed here. First, it is often important to include a variety of random effects in a model. For example, both CH and UH might be included (see below: Section 6.5). One flexible method for the inclusion of such terms is to include a log-linear term with additive random effects. This is often termed a convolution model. Besag et al. (1991) first suggested, for tract count effects, a rate parametrization of the form,

$$\exp\{x_i'\beta + u_i + v_i\},$$

where $x_i'\beta$ is a trend or fixed covariate component, u_i and v_i are correlated and uncorrelated heterogeneity, respectively. These components then have separate prior distributions. Often the correlated component is considered to have either an intrinsic Gaussian (CAR) prior distribution or a fully specified multivariate normal prior distribution.

6.4.1 Conditional autoregressive (CAR) models

6.4.1.1 Improper CAR (ICAR) models

The intrinsic autoregressions improper difference prior distribution, uses the definition of spatial distribution in terms of differences and allows the use of a singular normal joint distribution. This was first proposed by Besag et al.

(1991). Hence, the joint prior distribution for $\{u\}$ is defined as

$$p(\mathbf{u}|r) \propto \frac{1}{r^{m/2}} \exp\left\{-\frac{1}{2r}\sum_i \sum_{j \in \delta_i}(u_i - u_j)^2\right\}, \tag{6.11}$$

where δ_i is a neighborhood of the i th tract. The neighborhood δ_i was assumed to be defined for the first neighbor only. Hence, this is an example of a Markov random field model (see e.g., Rue and Held, 2005). More general weighting schemes could be used. For example neighborhoods could consist of first and second neighbors (defined by common boundary) or by a distance cut-off (for example, a region is a neighbor if the centroid is within a certain distance of the region in question). The uncorrelated heterogeneity (v_i) was defined by Besag et al. (1991) to have a conventional zero-mean Gaussian prior distribution:

$$p(v) \propto \sigma^{-m/2} \exp\left\{-\frac{1}{2\sigma}\sum_{i=1}^m v_i^2\right\}. \tag{6.12}$$

Both r and σ were assumed by Besag et al. (1991) to have improper inverse exponential hyperpriors:

$$\text{prior}(r, \sigma) \propto e^{-\epsilon/2r}e^{-\epsilon/2\sigma}, \qquad \sigma, r > 0, \tag{6.13}$$

where ϵ was taken as 0.001. These prior distributions penalize the absorbing state at zero, but provide considerable indifference over a large range. Alternative hyperpriors for these parameters which are now commonly used are in the gamma and inverse gamma family, which can be defined to penalize at zero but yield considerable uniformity over a wide range. In addition, these types of hyperpriors can also provide peaked distributions if required.

The posterior distribution derived from this set up can be sampled using McMC algorithms such as the Gibbs or Metropolis–Hastings samplers. A Gibbs sampler was used in the original example, as conditional distributions for the parameters were available in that formulation.

An advantage of the intrinsic Gaussian formulation is that the conditional moments are defined as simple functions of the neighboring values and number of neighbors $(n_{\delta i})$. and the conditional distribution is defined as:

$$[u_i|\dots] \sim N(\overline{u_i}, r/n_{\delta i}),$$

where $\overline{u_i} = \sum_{j \in \delta_i} u_j/n_{\delta i}$, the average over the neighborhood of the i th region.

6.4.1.2   Proper CAR (PCAR) models

While the intrinsic CAR model introduced above is useful in defining a correlated heterogeneity prior distribution, this is not the only specification of a Gaussian Markov random field (GMRF) model available. In fact, the improper CAR is a special case of a more general formulation where neighborhood dependence is admitted but which allows an additional correlation parameter (Stern and Cressie, 1999). Define the spatially-referenced vector of interest as $\{u_i\}$. One conditional specification of the proper CAR formulation yields:

$$[u_i | \ldots] \sim N(\mu_i, r/n_{\delta i}) \tag{6.14}$$

$$\mu_i = t_i + \phi \sum_{j \in \delta_i} (u_j - t_j)/n_{\delta_i} \tag{6.15}$$

where t_i is the trend $(=x_i'\beta)$, r is the variance, and ϕ is a correlation parameter. It can be shown that to ensure definiteness of the covariance matrix, ϕ must lie on a predefined range which is a function of the eigenvalues of a matrix. In detail, the range is the smallest and largest eigenvalues ($\phi_{\min} = \eta_1^{-1}, \phi_{\max} = \eta_m^{-1}$) of $diag\{n_{\delta_i}^{-1/2}\}.C.diag\{n_{\delta_i}^{1/2}\}$ where $C_{ij} = c_{ij}$, i.e., ($\phi_{\min} < \phi < \phi_{\max}$) and

$$c_{ij} = \begin{cases} \frac{1}{n_{\delta_i}} & \text{if } i \sim j \\ 0 & \text{otherwise} \end{cases} .$$

Of course, $\phi_{\min}$ and $\phi_{\max}$ can be precomputed before using the proper CAR as a prior distribution. It could simply be assumed that a (hyper) prior distribution for ϕ is $U(\phi_{\min}, \phi_{\max})$. As noted by Stern and Cressie (1999) this specification does lead to a simple form for the partial correlation between different sites. Note that in the simple case of no trend ($t_i = 0$) then the model reduces to

$$[u_i | \ldots] \sim N(\mu_i, r/n_{\delta i}) \tag{6.16}$$

$$\mu_i = \phi \bar{u}_i. \tag{6.17}$$

The main advantages of this model formulation is that it more closely mimics fully specified Gaussian covariance models, as it has a variance and correlation parameter specified, does not require matrix inversion within sampling algorithms, and can also be used as a data likelihood. Note that the ICAR and PCAR are available on OpenBUGS, nimble, and INLA. STAN also has ICAR models(see e.g., package brms)

6.4.1.3 Gaussian process convolution (PC) and GCM models

Higdon (2002) and Calder et al. (2002), (see also Calder, 2007, 2008, and Cressie and Wikle, 2011, ch. 4) suggested the use of kernel convolution models to discretely approximate continuous spatial Gaussian fields. The basic idea is that at location s the field is given by

$$u(s) = \sum_{j=1}^{M} k(w_j - s)x(w_j)$$

where $k(w_j - s)$ is a distance kernel, $\{w_j\}$ a set of evaluation points, and independent noise $x(w_j) \overset{iid}{\sim} N(0, \tau_u)$. This is a smoothing of white noise which leads to a correlated effect. Usually the kernel is assumed to be symmetric bivariate Gaussian and it can be pre-computed and so the main parameter controlling the field is τ_u, and the evaluation mesh. Higdon (2002) noted that for the continuous analogue, the given kernel has a one-to-one relation to the covariance of the resulting Gaussian field. This allows for great flexibility and the idea of *reduced rank* models has led to predictive process models (Banerjee et al., 2008).

Recent applications of the PC approach to parsimonious modeling of correlated spatial noise is found in Onicescu et al. (2017a) and Miller et al. (2020). A closely related model is the Gaussian component mixtures (GCM) which use neighborhood adjacencies instead of a distance kernel:

$$u_i = \frac{1}{w_i} \sum_{j=1}^{n_{\delta_i}} w_{ij} x_j$$

$$\text{with } \ x_j \overset{iid}{\sim} N(0, \tau_u)$$

and $w_{ij} = 1$ within the i th neighborhood and $w_i = \sum_{j=1}^{n_{\delta_i}} w_{ij}$.

Hence, for these weights, $u_i = \frac{1}{n_{\delta_i}} \sum_{j=1}^{n_{\delta_i}} x_j$, the neighborhood average. The resulting GCM does seem to approximate ICAR effects well, as demonstrated by Moraga and Lawson (2012). Note that the spatial model assumed in multilevel modeling closely resembles the GCM (see e.g., Leyland and Goldstein, 2001, p. 143-157) but includes $n_{\delta_i} + 1$ white noise terms.

6.4.1.4   Case event models

For case event data, where a point process model is appropriate, it is still possible to consider a form of log Gaussian Cox process where the intensity of the process is governed by a Spatial Gaussian process and conditional on the intensity the case distribution is a Poisson process. As an approximation to the Gaussian process a CAR prior distribution can be proposed. For example, define the first order intensity of the case events as

$$\lambda(s) = \lambda_0(s) \exp\{\beta + S(s)\}$$

where $S(s)$ is the Gaussian process component. At a given case location, s_i, this will yield a likelihood contribution

$$\lambda(s_i) = \lambda_0(s_i) \exp\{\beta + S(s_i)\}.$$

By considering an intrinsic Gaussian specification for $S(s_i)$ we can proceed by assuming that the prior distribution for $\{S(s_i)\}$ is a conditional autoregressive specification, i.e., for short, define $S_i \equiv S(s_i)$, and hence

$$[S_i | \ldots] \sim N(\overline{S}_{\delta_i}, r/n_{\delta i}).$$

where $\overline{S}_{\delta_i}$ is the mean of the S values in the neighborhood of S_i. This idea relies on the definition of a neighborhood. For case events, the definition of a neighborhood is problematic. Unlike polygonal regions there is not a simple definition of neighboring points. One possibility, would be to define a circular radius around each event and to include all points as neighbors if they fell within the radius. Of course, the size of the radius is arbitrary and hence the neighborhoods could be arbitrary also. An alternative would be to define a neighborhood by tesselation. Tesselation is a form of tiling which uniquely divides the space of points into areas. Usually, these tilings describe the area closest to the point, and to no other. In that sense they define "territories." Defining the neighborhood in this way leads to a Voronoi/Dirichlet tesselation. This tiling leads to sets of neighbors defined by the adjoining edges of the tile. An adjoining neighbor is known as a *natural neighbor*. The tiling defines *natural neighborhoods*. It is possible to define first (or greater) order neighbors in this way. On R the package deldir can be modified for this task. The dual of the Voronoi tesselation is the Delauney triangulation. This triangulation is formed by the bisectors of the tile edges. Each point will also have a number of Delauney neighbors. These are often the same as the Voronoi neighbors but they can also differ. One advantage of the Delauney neighbors is that the triangulation always forms a convex hull in the points, and to some degree suffers less from edge effects than the Voronoi tesselation.

Hence a hierarchical model can be specified with the i th likelihood contribution $\lambda(s_i) = \lambda_0(s_i) \exp\{\eta_i + v_i + S_i\}$ where $\eta_i = x_i'\beta$ a linear predictor with fixed covariate vector x_i', and

$$[S_i | \ldots] \sim N(\overline{S}_{\delta_i}, r/n_{\delta i})$$
$$v_i \sim N(0, \kappa_v)$$
$$\beta \sim \mathbf{N}(\mathbf{0}, \mathbf{\Gamma}_\beta).$$

Of course, in this formulation, then both $\lambda_0(s_i)$ must be estimated, and also the integral of the intensity must be computed. An example of this type of analysis is given in Hossain and Lawson (2008). The conditional logistic likelihood model (6.6) can be fitted if a control disease is available, and this obviates the necessity of estimating $\lambda_0(s_i)$. However the specification of the spatial structure is different as the joint distribution of cases and controls is considered under the conditional model.

6.4.2   Fully specified covariance models

An alternative specification involves only one random effect for both CH and UH. This can be achieved by specifying a prior distribution having two parameters governing these effects. For example, the covariance matrix of an MVN prior distribution can be parametrically modelled with such terms (Diggle et al., 1998; Wikle, 2002). This approach is akin to universal Kriging (Wackernagel, 2003; Cressie, 1993), which employs covariance models including variance and covariance range parameters. It has been dubbed "*generalized linear spatial modeling.*" A software library is available in R (geoRglm). Usually, these parameters define a multiplicative relation between CH and UH. For the full Bayesian analysis of this model, use is often made of posterior sampling algorithms.

In the parametric approach of Diggle et al. (1998), which was originally specified for point process models, the first-order intensity of the process was specified as

$$\lambda(s) = \lambda_0(s) \exp\{\beta + S(s)\}, \tag{6.18}$$

where β is a non-zero mean level of the process, and $S(s)$ is a zero mean Gaussian process with, for example, a powered exponential correlation function defined for the distance d_{ij} between the i th and j th locations as $\rho(d_{ij}) = \exp\{-(d_{ij}/\phi)^\kappa\}$ and variance σ^2. Other forms of covariance function can be specified. One popular example is the Matérn class defined for the distance (d_{ij}) as

$$\rho(d_{ij}) = (d_{ij}/\phi)^\kappa K_\kappa(d_{ij}/\phi)/[2^{\kappa-1}\Gamma(\kappa)] \tag{6.19}$$

where $K_\kappa(.)$ is a modified Bessel function of the third kind. In this case, the parameter vector $\theta = (\beta, \sigma, \phi, \kappa)$ is updated via a Metropolis–Hastings-like (Langevin-Hastings) step, followed by pointwise updating of the S surface. Conditional simulation of S surface values at arbitrary spatial locations (non-data locations) can be achieved by inclusion of an additional step once the sampler has converged. Covariates can be included in this formulation in a variety of ways. For count data, the equivalent Poisson mean specification could be

$$\mu_i = e_i \exp\{\beta + S_i\},$$

where $\mathbf{S} \sim \mathbf{MVN}(\mathbf{0}, \mathbf{\Gamma})$ and $\mathbf{\Gamma}$ is a spatial covariance matrix (Kelsall and Wakefield, 2002). In comparisons of CAR and fully-specified covariance models there appears to be different conclusions about which are more useful in recovering relative risk in disease maps (Best et al., 2005, Henderson et al., 2002). Note that when the above formulation is applied to a point process then a log Gaussian Cox process (LGCP) results . There is now a range of software available for fitting such models: the R package lgcp provides McMC implementation and the stochastic partial differential equation (SPDE) approach on a finite element mesh is available in INLA (Illian et al., 2012). Comparison of these approaches is given in Taylor and Diggle (2014) and Taylor et al. (2015).

6.5 Convolution Models

Often it is important to employ both CH and UH random effects within the specification of η_i. The rationale for this lies in the basic assumption that unobserved effects within a study area could take on a variety of forms. It is always prudent to include a UH effect to allow for uncorrelated extra variation. However, without prior knowledge of the unobserved confounding, there is no reason to exclude either effect from the analysis and it is simple to include both effects within an additive model formulation such as $\eta_i = v_i + u_i$. In general, these two random effects are not well identified, although the assumption of a correlated prior distribution for u_i should provide a degree of identification between UH and CH. Given that we are usually interested in the total effect of unobserved confounding then the sum of the effects is the important component and that is well identified. Discussion of these identifiability issues is given in Eberley and Carlin (2000). Occasionally, the computation of the relative variance contribution (intraclass correlation) can be useful. If the variance of the UH and CH component are, respectively, κ_v and κ_u, then this is just given by $\frac{\kappa_v}{\kappa_v + \kappa_u}$. Of course if these components are not identified then this computation will not be useful.

Convolution models, or as they are also known: BYM models, with ICAR CH random effects, have been shown to be robust under simulation to a wide range of underlying true risk models (Lawson et al., 2000) and so are widely used for the analysis of relative risk in disease mapping.

6.5.1 Leroux prior specification

The BYM convolution model is defined by $\exp\{\alpha_0 + u_i + v_i\}$, with suitable CH prior distribution for u_i and UH prior distribution for v_i and an intercept α_0. The lack of identification in this set up has led to an alternative proposal that has now been implemented in various packages: the Leroux model. Leroux et al. (2000) first proposed a model whereby the covariance structure of the model is a mixture between an uncorrelated and correlated effect. More formally

$$\mathbf{u} \sim \mathbf{N}(\mathbf{0}, \tau^2 Q(W, \rho)^{-1}),$$

where the precision matrix is defined as $Q(W, \rho) = \rho[diag(W1) - W] + (1 - \rho)I$ and W is a neighborhood weight matrix. In this formulation the ρ parameter controls the overall degree to which the effect is correlated: $\rho = 1$ gives an ICAR model whereas $\rho = 0$ gives an uncorrelated (UH) effect. Hence the model allows the degree of smoothing or clustering to be estimated, without having to use a convolution with two components. Congdon (2010) ch. 4 provides discussion of variants of the Leroux model. If one is simply concerned that confounding effects be estimated without interest in separate components then the convolution or Leroux models can be used. If separate effects are of interest then the Leroux model does not provide estimates of these. Of course the convolution model may suffer from identifiability. A recent simulation-based comparison of the convolution and Leroux models can be found in Baer and Lawson (2019). The Leroux model is implemented on CARBayes, can be programmed on Win/OpenBUGS, and also is available on INLA (leroux.inla).

Part I

Basic Software Approaches

7

BRugs/OpenBUGS

The now traditional vehicle for the analysis of BHMs and in particular Bayesian disease mapping (BDM) models is WinBUGS (https://www.mrc-bsu.cam.ac.uk/software/bugs/the-bugs-project-winbugs). WinBUGS is a windows package that was originally designed to estimate BHMs via Gibbs Sampling and later incorporated MH and variant samplers. The last stable version was 1.4.3 in 2007. A later development of an open source software version called OpenBUGS has essentially the same programming style but adds some extra features to the computational and presentational palette (version 3.2.3 from 2014). A parrallelizable variant of OpenBUGS, called MultiBUGS, is currently under development (version 1.0.0 in 2018). On R it is possible to both invoke WinBUGS via the package R2WinBUGS (or its variant R2OpenBUGS) or to directly invoke BUGS code using the package BRugs. BRugs uses OpenBUGS but does not use the windowed procedures common to R2Win/OpenBUGS and so here we will use BRugs for subsequent exposition.

7.1 Uncorrelated Heterogeneity Models

7.1.1 A BUGS model: The gamma-Poisson model

The first important step is the specification of a model. I will demonstrate a simple DM model to provide a start point for development. The following is a gamma-Poisson spatial model which will be fitted to the SC respiratory cancers dataset.

The model is

$$y_i \sim Pois(e_i\theta_i)$$
$$\theta_i \sim gamma(a, b)$$
$$a \sim Exp(\upsilon)$$
$$b \sim Exp(\rho)$$

The following is the model code that describes this hierarchy:

```
model{
for(i in 1:m){
Y1998[i]~dpois(mu[i])          #1
mu[i]<-Exp98[i]*theta[i]
theta[i]~dgamma(a,b)      #2
}
a~dexp(10)       #3
b~dexp(10)       #4
}
```

The main model code is specified after the word "model" and within braces: {}.

Here the outcome is the count of respiratory cancers in 1998 in SC counties (Y1998[i]) with the expected count (Exp98[i]). The relative risk parameter in each county is theta[i]. The first line of the for loop (#1) represents the data model (Poisson likelihood), while line #2 represents the gamma prior distribution for θ_i. Each of these items must be assigned for each of the regions (m = 46 SC counties) and so must appear inside the for loop.

The prior distributions for the other parameters are outside the for loop as they are not region specific (a,b). Note that distributions are assigned using the "~" symbol and usually start with the letter "d": dpois, dgamma, dexp etc.

This code is then held in a text file for use with BRugs. To process such a model a number of steps are needed. First the model code must be in a text file. Second the data must be in list format within a text file. Third, the initial values for the sampler must be in a text file also in list format. The following sequence of commands checks the correctness of the model, loads the data, compiles the model with 2 chains, and initializes both chains with inits from the file "inits.txt."

The data file ("data.txt") looks like this:

```
list(m=46,Y1998=c( 18,90,10,120,12,14,76,96,10,256,37,23,40,29,36,
55,21,63,15,19,129,47,260,60,10,184,22,45,43,44,10,171,11,
34,22,34,51,63,90,201,10,202,87,25,25,91 ),
Exp98=c(19.334642001146, 105.221991510865, 8.9954123633133,
126.211287025262,
12.9499400671852, 17.0850039703209, 85.5262771111914, 107.178846922884,
11.0291918950188, 248.419380066852, 38.5954996425929, 27.0027208298727,
```

```
13.3636034454982, 194.239681727817, 84.0882670370562, 23.9367452023769,
29.1377576211652, 121.126445726))
```

While the inits file ("inits.txt") looks like:

```
list(a=10,b=10,theta=c(1, 1, 1, 1, 1, 1, 1, 1, 1, 1, 1, 1, 1, 1, 1, 1, 1, 1,
1, 1,
     1, 1, 1, 1, 1, 1, 1, 1, 1, 1, 1, 1, 1, 1, 1, 1, 1, 1, 1, 1, 1,
     1, 1, 1, 1, 1))
```

On BRugs the command sequence is:

```
modelCheck("modelfile.txt")
modelData("data.txt")
modelCompile(numChains=2)
modelInits(rep("inits.txt",2))
```

Once these commands are successfully performed the model is ready for updating. The command modelUpdate can be used:

```
modelUpdate(10000)
```

This will run the model for 10000 iterations. To check the model samples the command sequence:

```
samplesSet(c("theta","mu","a","b","deviance"))
dicSet()
modelUpdate(2000)
dicStats()
asd<-samplesStats("*")
```

will set the DIC and monitor for parameters theta, mu, a, b, deviance, and deliver the DIC summary and posterior mean and credible intervals of the monitored parameters. The object asd contains the sample summary statistics in the form of mean, sd, MC error, 2.5%ile, median, 97.5%ile. Note that if the deviance has been monitored then the DIC using the alternate pV measure can be computed easily as

$$DIC_v = mean(\text{deviance}) + sd(\text{deviance})^2/2.$$

Out from the samplesStats command for this example looks like this:

```
head(samplesStats("*"))
```

	mean	sd	MC_error	val2.5pc	median	val97.5pc	start	sample
a	1.339	0.2482	0.008576	0.9172	1.319	1.888	10001	4000
b	1.095	0.2432	0.008388	0.673	1.075	1.64	10001	4000
deviance	301	9.259	0.1595	284.9	300.2	320.7	10001	4000
mu[1]	18.38	4.229	0.06581	11	18.03	27.59	10001	4000
mu[2]	90.46	9.54	0.1468	72.7	90.2	109.8	10001	4000
mu[3]	10.1	2.977	0.04838	5.172	9.76	16.63	10001	4000

FIGURE 7.1
Sample monitor output from BRugs

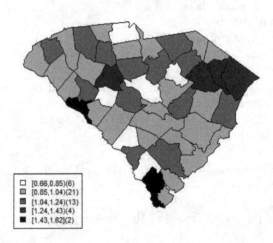

FIGURE 7.2
Posterior mean relative risk map for the gamma-Poisson model applied to the SC county respiratory cancers data.

As the model produces posterior mean estimates of the relative risk (θ_i) we can map these using thematic mapping tools. Figure 7.2 displays the posterior mean estimates of theta for the SC county respiratory cancers example.

Figure 7.3 displays the standard deviation estimated from the posterior sample.

7.1.2   Log-normal models

The gamma-Poisson is not easily extendable to include correlation effects and so it is often preferable to initially consider a log-normal formulation. In that case the relative risk is linked to a predictor (usually linear) which can include a range of additive and non-additive effects. The likelihood model assumed

relative risk SD

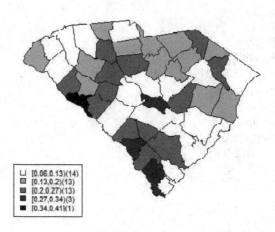

FIGURE 7.3
SD map of the relative risk for the gamma-Poisson model for SC county respiratory cancers.

for this example is Poisson with $y_i \sim Poisson(e_i\theta_i)$ and a basic model is assumed $\eta_i = \beta_0 + z_{1i}$.

Here the extra-variation is modeled as uncorrelated heterogeneity (UH) with a zero mean Gaussian prior distribution, i.e., $z_{1i} \sim N(0, \tau_{z_1})$. The hierarchy is then

$$y_i \sim Poisson(e_i\theta_i)$$
$$\log(\theta_i) = \beta_0 + z_{1i}$$
$$\beta_0 \sim N(0, \tau_\beta)$$
$$z_{1i} \sim N(0, \tau_{z_1})$$
$$\tau_*^{-1} \sim gamma(2, 0.5)$$

Both the intercept and UH term have variances, with gamma prior distributions for the precisions. The choice of prior distributions for precisions is important particularly when random effects are included in models. Fixed values for precisions can lead to difficulties with convergence due to early truncation of the hierarchy.

The model code is given below:

```
model{
for(i in 1:m){
Y1998[i]~dpois(mu[i])
mu[i]<-Exp98[i]*theta[i]
log(theta[i])<-a0+z1[i]
z1[i]~dnorm(0,tauZ)
}
a0~dnorm(0,tau0)
tauZ~dgamma(2,0.5)
tau0~dgamma(2,0.5)
}
```

Run code is:

```
modelCheck("log_normal_BUGS_model.txt")
modelData("BUGS_data.txt")
modelCompile(numChains=2)
modelInits(rep("log_normal_BUGS_inits.txt",2))
modelUpdate(10000)
samplesSet(c("theta","a0","z1","deviance"))
dicSet()
modelUpdate(2000)
dicStats()
asd<-samplesStats("*")
library(coda)
gelman.plot(buildMCMC("deviance"))
```

Figure 7.4 displays the BGR plot for the log normal (UH) model for the SC county level respiratory cancers data example. Figure 7.5 displays the posterior mean UH effect for the basic log-normal model.

Figure 7.6 displays the posterior mean relative risk estimates for the log-normal UH model

7.2 Convolution models

Both the gamma-Poisson and the log normal models address extra variation in the count incidence data but do not explicitly address spatial correlation. Besag et al. (1991) first suggested the use of an extension of the log-normal

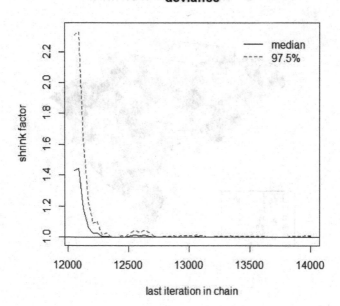

FIGURE 7.4
BGR plot of deviance for the log-normal model for the SC respiratory cancers example.

model to incorporate a spatially structured component. In the original model, they added a component random effect which was assumed to have a prior distribution incorporating spatial correlation. The convolution hierarchy is assumed to be

$$y_i \sim Poisson(e_i\theta_i)$$
$$\log(\theta_i) = \beta_0 + v_i + u_i$$
$$\beta_0 \sim N(0, \tau_\beta)$$
$$v_i \sim N(0, \tau_v)$$
$$u_i|\mathbf{u}_{-i} \sim N(\overline{u}_{\delta_i}, \tau_u/n_{\delta_i})$$
$$\tau_*^{-1} \sim gamma(2, 0.5)$$

$N(\overline{u}_{\delta_i}, \tau_u/n_{\delta_i})$ is known as an ICAR model and is a singular normal distribution conditionally defined on the other components. For shorthand this can be written $u_i \sim ICAR(\tau_u)$. On WinBUGS and OpenBUGS the ICAR model is defined as a multivariate node with the car.normal distribution. The

UH effect estimate: LN model

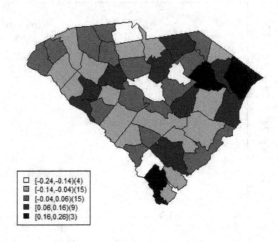

FIGURE 7.5
Posterior mean uncorrelated effect from the basic log normal model for SC
respiratory cancers.

distribution relies on the specification of adjacencies between regions. These
are held in a sparse adjacency vector. A typical call to car.normal could be

```
u[1:m]~car.normal(adj[],wei[],num[],tauU)
for(k in 1:sumNumNeigh) {wei[k] <- 1 }
```

Here, the u vector has associated with it an adjacency vector (adj), a number of neighbors vector (num), and a weight vector (wei). The distribution is
controlled by the precision tauU. The adjacency vector completely character-
izes the spatial structure and is usually only specified by adjoining neighbors:
i.e., if two regions have a common boundary then they are neighbors otherwise
they are not. On Win/OpenBUGS the adj and num vectors can be obtained
from the adjacency tool in GeoBUGS based on a GeoBUGS format *.map
file. Adjacencies can be obtained directly from shapefiles on R also and this is
discussed in Chapter 2. sumNumNeigh is the sum of all the neighbors of each
region and can be computed by the command sum(num). The weight vector

theta estimate: LN model

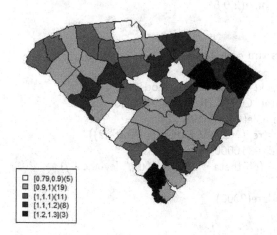

[0.79,0.9)(5)
[0.9,1)(19)
[1,1.1)(11)
[1.1,1.2)(8)
[1.2,1.3)(3)

FIGURE 7.6
Relative risk posterior mean estimates for the log-normal UH model for the SC county level respiratory cancers data.

is of length sumNumNeigh and is available for specification of inter-neighbor dependencies. This could be relevant if it is thought that different neighbors can interact differently with a given region. Often it is simply set to the value 1 if the adjacencies alone will determine the spatial dependence. The model code is given below:

```
model{
 for(i in 1:m){
 Y1998[i]~dpois(mu[i])
 mu[i]<-Exp98[i]*theta[i]
 log(theta[i])<-a0+v[i]+u[i]
 v[i]~dnorm(0,tauV)}
 u[1:m]~car.normal(adj[],wei[],num[],tauU)
 for(k in 1:sumNumNeigh) {
 wei[k] <- 1 }
 a0~dnorm(0,tau0)
```

```
tauV~dgamma(2,0.5)
tau0~dgamma(2,0.5)
tauU~dgamma(2,0.5)
  }
```

The BRugs run code is:

```
modelCheck("CONV_BUGS_model.txt")
modelData("CONV_BUGS_data.txt")
modelCompile(numChains=2)
modelInits(rep("CONV_BUGS_inits.txt",2))
modelUpdate(10000)
samplesSet(c("theta","a0","v","u","deviance"))
dicSet()
modelUpdate(2000)
dicStats()
asd<-samplesStats("*")
library(coda)
gelman.plot(buildMCMC("deviance"))
```

Figures 7.7, 7.8, and 7.9 display the posterior mean estimates of the three components: θ_i, v_i, u_i for the convolution model described above, applied to the SC respiratory cancers data. The model was run for 10000 burn-in and 2000 sample size. The BGR plot for the deviance in the final sample is displayed in Figure 7.10.

7.3   Goodness of Fit

The above examples have associated with them goodness of fit measures and the classic measure available on BRugs is the DIC (see Chapter 5). This can be obtained, with a sample size of 2000, using:

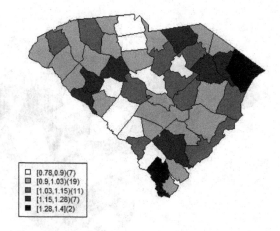

FIGURE 7.7
Posterior mean of the relative risk (θ_i) under the convolution model.

```
dicSet()
modelUpdate(2000)
dicStats()
```

For the above three models for the SC respiratory cancers example the DIC and pD (effective number of parameters) is:

Model	DIC	pD
Gamma-Poisson	345.1	44.13
log-normal	323.7	29.77
convolution	317.3	21.73

Based on the relative comparison of DICs, this suggests that the convolution model yields the best fit for these data. It is also has the lowest pD, which speaks to the parsimony of this model.

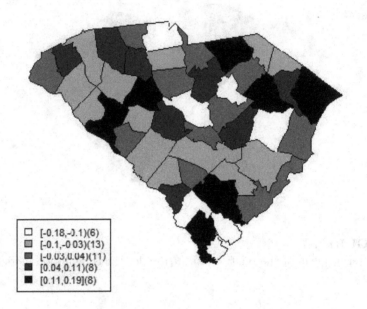

FIGURE 7.8
Posterior mean of the uncorrelated component (v_i) for the convolution model
for the SC respiratory cancers data.

7.4   Model variants

7.4.1   Simple mixtures

One extension of the convolution model is to allow the components within
the model to vary with location. This variant assumes that the there are two
random effects but that they vary in their importance across the study region.

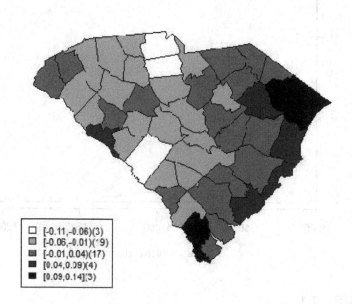

FIGURE 7.9
Posterior mean of the correlated ICAR component (u_i) for the convolution model for the SC respiratory cancers data.

In this case you can assume:

$$y_i \sim Poisson(e_i\theta_i)$$
$$\log(\theta_i) = \beta_0 + p_i v_i + (1 - p_i)u_i$$
$$p_i \sim beta(1,1)$$

The prior distribution for p_i is here assumed to be uniform across $(0,1)$: $p_i \sim beta(1,1)$. Figure 7.11 displays the resulting posterior mean estimates for p_i, v_i, u_i for the SC respiratory cancers example.

Alternatively the logit of the probability could be modeled directly via a linear or non-linear predictor, e.g.,

$$\log it(p_i) = \alpha_0 + f(x_i^t \alpha)$$

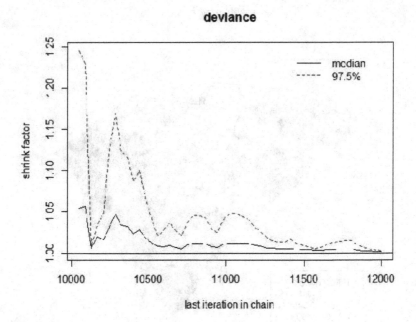

FIGURE 7.10
gelman.plot (BGR) of the deviance for the final sample for the convolution
model.

and this would provide a mechanism for directly modeling clustering related
to predictors in **x**.

7.4.2   Leroux model

An alternative to a convolution model which does not include *separate* random
effect components is the Leroux model (see Section 6.5.1). This model allows
the inclusion of a scale estimate of the degree to which the data are correlated.
One version of the code uses the conditional mean and variance to model the
effect MacNab et al. (2006). However, this does not sample the joint posterior
distribution correctly and should be regarded as an approximation only. The
code below directly evaluates the multivariate normal prior distribution based
on the extended precision matrix: $Q(W, \rho) = \rho[diag(W1) - W] + (1 - \rho)I$. Note
that the binary adjacency matrix W is fixed and is scaled by parameter ρ.
Hence the sampling required is simply for ρ and the precision parameter τ^{-1}.
In the code the precision is prec and rho is the "degree of spatial structure"
parameter. For the code the matrix Q1, identity matrix RD, and zero vector
R0 are read in.

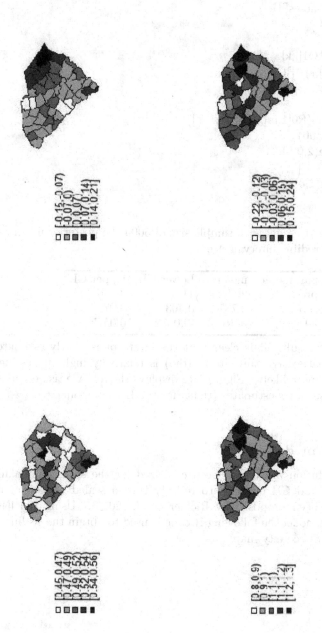

FIGURE 7.11

SMix posterior mean estimates, row-wise left to right: mixing probability, CH component, relative risk, UH component.

```
model
{for (i in 1:m){Y1998[i]~dpois(mu[i])
log(mu[i])<-log(Exp98[i])+log(theta[i])
log(theta[i])<-a0+s[i]}
for (j in 1:m){
for(k in 1:m){
asd[j,k]<-rho*Q1[j,k]
aer[j,k]<-asd[j,k]+RD[j,k]
adf[j,k]<-prec*aer[j,k]
}}
s[1:m]~dmnorm(R0[1:m],adf[1:m,1:m])
a0~dnorm(0,tau0)
tau0~dgamma(2,0.5)
rho~dunif(0,1)
prec~dgamma(1,0.5)
}
```

For the converged model with sample size of 5000 the posterior mean estimates and 95% credible intervals are:

parameter	mean	Lower CL	Upper CL
prec	20.52	11.05	34.10
rho	0.787	0.463	0.979
a0	-0.0113	-0.0772	0.0575

Based on these results, it is clear that the intercept is poorly estimated, the "degree of spatial structure" term (rho) is relatively high suggesting a clustered effect is more likely. Figure 7.12 displays the relative risk estimates (left) and the posterior *s* estimates (right) for the Leroux model.

7.4.3   BYM2 model

The BYM convolution model has been criticized for the scaling imbalance between the UH and CH effects. To remedy this, a scaled version of the ICAR effect has been proposed by Riebler et al. (2016). Using the INLA function inla.scale.model the following R code is used to obtain the scaling, in this case, for the SC county map:

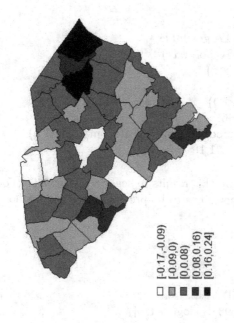

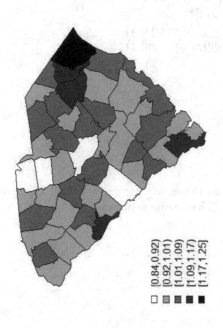

FIGURE 7.12

Posterior mean estimate maps of θ_i and s_i for the Leroux model based on the full MVN prior distribution on BRugs

```
library(INLA)
#SC_poly.txt is a INLA graph file
g=inla.read.graph("SC_poly.txt")
Q=-inla.graph2matrix(g)
diag(Q) = 0
diag(Q) = -rowSums(Q)
n=dim(Q)[1] # n=46
Q.scaled=inla.scale.model(Q,constr=list(A=matrix(1,1,n),e=0))
scale=Q.scaled[1,1]/Q[1,1]
```

For the SC county map this results in scale = 0.3783966. For running on BRugs, the scale parameter has to be pre-computed. The following code fits the BYM2 model:

```
model{
scale<-0.3783966
for(i in 1: m){
Y1998[i]~dpois(mu[i])
log(mu[i])<-log(Exp98[i])+log(theta[i])
theta[i]<-exp(ac0+bym2[i])
vc[i]~dnorm(0,1)
Corr[i]<-sigma*uc[i]*sqrt(rho/scale)
UCorr[i]<-sigma*vc[i]*sqrt((1-rho))
bym2[i]<-Corr[i]+UCorr[i]}
for (k in 1 :sumNumNeigh){wei[k]<-1}
uc[1:m]~car.normal(adj[],wei[],num[],1)
sigma~dunif(0,5)
tau<-pow(sigma,-2)
rho~dbeta(1,1)
ac0~dnorm(0,tau0)
tau0<-pow(sd0,-2)
sd0~dunif(0,5)
})
```

The model for these data does not converge within 10K burn-in, and required 50K for stability. The results reported here are for a sample of 5000 after a 50K burn-in.

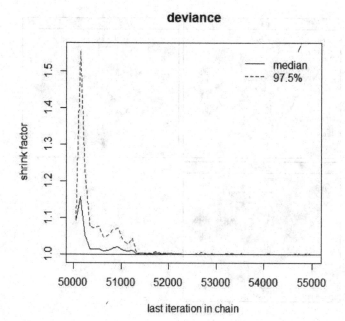

FIGURE 7.13
BGR plot for the deviance for the BYM2 model fitted to the SC county
respiratory cancers data.

The BGR plot for the sample of the deviance is displayed in Figure 7.13.
The posterior mean estimates for UCorr, Corr, and relative risk are displayed
in Figure 7.14. The scaling effect can be seen in the range of values of the
Corr and UCorr components.

7.4.4 DIC comparisons

It is notable that there are a range of DIC measures for the model variants that
have been presented here. The DIC for the Leroux model is 319.1 (pD=23.66)
and for the spatial mixture 299.6 (pD=5.50). For the BYM2 model the DIC is
300.7 (pD = 1.62). Of these models the spatial mixture appears to provide the
lowest DIC and pD, although the BYM2 model is very close and has a small
pD. However this also implies that the deviance for the BYM2 model is larger
than that of the spatial mixture. Of course, different model formulations could
be preferred, based on other grounds besides empirical goodness of fit.

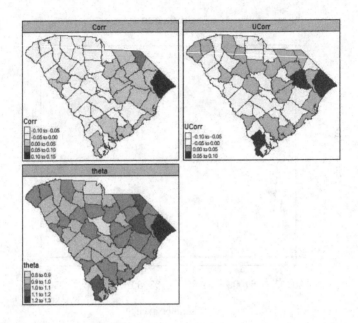

FIGURE 7.14
Posterior mean estimates of the components of the BYM2 model for the SC
county level respiratory cancers example.

8

Nimble

A recent development within the R world is the arrival of nimble. nimble is a wide ranging package which can parse BUGS or JAGS code and to write its own McMC samplers. To quote the package description:

"A system for writing hierarchical statistical models largely compatible with 'BUGS' and 'JAGS', writing nimbleFunctions to operate models and do basic R-style math, and compiling both models and nimbleFunctions via custom-generated C++. 'NIMBLE' includes default methods for MCMC, particle filtering, Monte Carlo Expectation Maximization, and some other tools. The nimbleFunction system makes it easy to do things like implement new MCMC samplers from R, customize the assignment of samplers to different parts of a model from R, and compile the new samplers automatically via C++ alongside the samplers 'NIMBLE' provides. 'NIMBLE' extends the 'BUGS'/'JAGS' language by making it extensible: New distributions and functions can be added, including as calls to external compiled code. Although most people think of MCMC as the main goal of the 'BUGS'/'JAGS' language for writing models, one can use 'NIMBLE' for writing arbitrary other kinds of model-generic algorithms as well."

At a basic level nimble can convert BUGS code almost directly into nimble models and then run the code via purpose written C++.

8.1 Gamma-Poisson Example

A simple example of the programming steps is a useful taster. I will run the gamma-Poisson model for the SC county level respiratory cancers example first. A useful reference to basic BDM using nimble is given in Lawson (2020). For any nimble model fit a number of steps have to be processed. As prerequisite the model code, data, constants, and inits have to be available in named lists. The code structure is essentially that of BUGS or JAGS. It is specified via the nimbleCode() statement. Outcome data are specified in the data list, and all other fixed inputs in the constants list. Finally all initial values are specied in the inits list. For the SC respiratory cancers example we would have:

```
GPdata<-list(Y1998=c( 18,90,10,120,12,14,76,96,10,256,37,23,40,29,36,
55,21,63,15,19,129,47,260,60,10,184,22,45,43,44,10,171,11,
34,22,34,51,63,90,201,10,202,87,25,25,91))
GPconsts<-list(m=46,Exp98=c(19.334642001146, 105.221991510865,
8.9954123633133, 126.211287025262,
12.9499400671852, 17.0850039703209, 85.5262771111914, 107.178846922884,

11.0291918950188, 248.419380066852, 38.5954996425929, 27.0027208298727,

32.2453350684913, 24.1871410613557, 29.3284980403873, 52.0933278275436,

23.3496100847714, 69.1791167378613, 15.7011547559647, 17.5779462883105,

98.0421453601469, 42.1724712080047, 277.747093167242, 49.9402374163248,

15.0708479385354, 137.177683720537, 13.3400552455942, 38.1425892644401,
46.222761591486, 49.646669857522, 16.011990994697, 161.116783742905,
7.49225226944375, 27.1667732892036, 23.2255895652772, 27.0506021696774,

50.282471254929, 68.9687528187193, 84.0568694371842, 241.020535657027,

13.3636034454982, 194.239681727817, 84.0882670370562, 23.9367452023769,

29.1377576211652, 121.126445726))


GPinits<-list(a=10,b=10,theta=c(1, 1, 1, 1, 1, 1, 1, 1, 1, 1, 1, 1, 1, 1, 1, 1, 1,
1, 1, 1, 1, 1, 1, 1, 1, 1, 1, 1, 1, 1, 1, 1, 1, 1, 1, 1, 1, 1, 1, 1, 1, 1, 1, 1, 1, 1))
```

The model code is as follows:

```
GPModel<-nimbleCode({
for(i in 1:m){
Y1998[i]~dpois(mu[i])
mu[i]<-Exp98[i]*theta[i]
theta[i]~dgamma(a,b)
}
a~dexp(10)
b~dexp(10)
})
```

Once these named lists and code are assembled, there are a number of setup steps to process and there are run options to consider.

Step 1) nimbleModel()

the first step is to assess the model code for consistency and create a NIMBLE model, which can be compiled and run. The format for a basic call is

Lmodel<-nimbleModel(code, constants = list(), data = list(), inits = list(),name="" ......).

For the GP example this could be

```
GPmod<-nimbleModel(code=GPmodel,name="GPM",constants=GPconsts,
data=GPdata,inits=GPinits)
```

output is :

```
defining model...
building model...
setting data and initial values...
running calculate on model (any error reports that follow may simply reflect
missing values in model variables) ...
checking model sizes and dimensions...
model building finished.
```

Step 2) Compilation

```
CGP<-compileNimble(GPM)
```

Step 3) Configure: Once a model is compiled it can then be configured for sampling and monitors added

```
CGPcon<-configureMCMC(CGP,print=TRUE)
CGPcon$addMonitors(c("a","b","theta"))
```

The output is a list of samplers that have been chosen

```
[1] RW sampler: a
[2] RW sampler: b
[3] conjugate_dgamma_dpois sampler: theta[1]
[4] conjugate_dgamma_dpois sampler: theta[2]
[5] conjugate_dgamma_dpois sampler: theta[3]
.
.
.
.
```

Step 4) The model can then be built and compiled for running:

```
GPMCMC<-buildMCMC(CGPcon)
CGPMCMC<-compileNimble(GPMCMC)
```

CGPMCMC is now ready for running. There are two methods for running such a compiled nimble object. The quick way is simply

```
CGPMCMC$run(niter=10000)
```

The more sophisticated way is to use runMCMC() which provides more control of the run in terms of iterations, burnin, thinning, and output style.

```
niter=10000
samples<-runMCMC(CGPMCMC,niter=niter,nburnin=9000,nchains=1,
summary=TRUE)
samples$summary
```

The above example runs for a burnin of 9000 iterations and samples the next 1000 with one chain and results are in object samples. samples$summary yields:

	Mean	Median	St.Dev.	95%CI_low	95%CI_upp
a	1.361761	1.314597	0.2486583	0.9987881	1.94559
b	1.108682	1.078349	0.2387562	0.7327587	1.671098

A complete set of commands for this example is given below, where monitors are added as well:

```
GPmod<-nimbleModel(code=GPModel,name="GPM",constants=GPconsts,
data=GPdata,inits=GPinits)
CGP<-compileNimble(GPmod)
CGPcon<-configureMCMC(CGP,print=TRUE)
CGPcon$addMonitors(c("a","b","theta"))
GPMCMC<-buildMCMC(CGPcon)
CGPMCMC<-compileNimble(GPMCMC)
niter=10000
samples<-runMCMC(CGPMCMC,niter=niter,nburnin=9000,nchains=1,
summary=TRUE)
samples$summary
```

For this example the summary yields posterior mean , median, sd, and 95% credible intervals for the monitored parameters:

	Mean	Median	St.Dev.	95%CI_low	95%CI_upp
a	1.4035638	1.3479619	0.31323206	0.941699	2.1477901
b	1.1535848	1.0874432	0.29761719	0.7149476	1.8440541
theta[1]	0.9442921	0.9305774	0.20375177	0.581418	1.372141
theta[2]	0.8627353	0.8570176	0.08946968	0.6936821	1.0364309
theta[3]	1.1338449	1.1023295	0.33093479	0.6054866	1.9031865
theta[4]	0.9531377	0.9535452	0.08913723	0.7818807	1.141926

.
.
.

samples\$summary is a matrix and relevant subsections can be downloaded and examined. For example, the posterior mean theta estimates can be accessed by

```
theta<-samples$summary[3:48,1]
```

These can then be mapped or otherwise processed. Figure 8.1 displays the posterior mean theta estimates for this nimble model fit.

Convergence of the McMC run can be checked using the CODA package and there is an option to specify that the output should be in CODA format. For example,

```
asd<-runMCMC(CGPMCMC,niter=niter,nburnin=9000,nchains=1,
summary=TRUE,samplesAsCodaMCMC=TRUE)
```

will yield a vector of the chain output(s) and also a summary in asd\$summary in matrix form. WAIC can also be computed automatically on nimble. For runMCMC the WAIC must be enabled in the configure step:

```
CGPcon<-configureMCMC(CGP,print=TRUE,enableWAIC=TRUE)
```

The WAIC is then available in the final sample output, in this case WAIC=329.1115:

```
asd<-runMCMC(CGPMCMC,niter=niter,nburnin=9000,nchains=1,
summary=TRUE,WAIC=TRUE)
```

An alternative short cut for running nimble models is provided by the nimbleMCMC() command. This subsumes the above steps within one command. For example

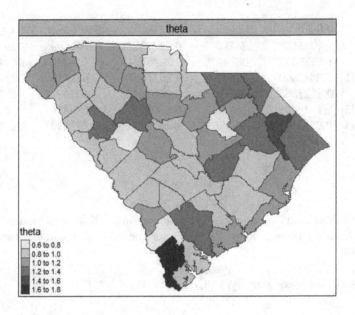

FIGURE 8.1
Relative risk (θ_i) posterior mean estimates for the gamma-Poisson model, for a sample of 1000, using the runMCMC command in nimble.

```
mcmc.out<-nimbleMCMC(code=GPModel,constants=GPconsts,
  data=GPdata,inits=GPinits,nchains=2,niter=10000,summary=TRUE,
WAIC=TRUE)
```

will fit the above model and also report the WAIC.

8.2 Log Normal Model

The log normal model is straightforward to program in nimble and uses the same data as the GP model. In this case the code is given as:

```
LNmodel<-nimbleCode({
  for(i in 1:m){
  Y1998[i]~dpois(mu[i])
```

```
mu[i]<-Exp98[i]*theta[i]
log(theta[i])<-a0+z1[i]
z1[i]~dnorm(0,tauZ)
}
a0~dnorm(0,tau0)
tauZ~dgamma(2,0.5)
tau0~dgamma(2,0.5)
})
```

The run code is:

```
LNmod<-nimbleModel(code=LNmodel,name="LNM",constants=LNconsts,
data=LNdata,inits=LNinits)
CLN<-compileNimble(LNmod)
CLNcon<-configureMCMC(CLN,print=TRUE,enableWAIC=TRUE)
CLNcon$addMonitors(c("a0","z1","theta"))
LNMCMC<-buildMCMC(CLNcon)
CLNMCMC<-compileNimble(LNMCMC)
niter=10000
samples<-runMCMC(CLNMCMC,niter=niter,nburnin=9000,nchains=1,
summary=TRUE,WAIC=TRUE)
```

The resulting WAIC is 313.36. Hence the LN model has a lower WAIC than the GP model. The sample summary yields the posterior mean estimates of z1 and theta. Figure 8.2 displays these estimates.

8.3 Convolution Model

The BYM or convolution model can also be fitted easily on nimble. Nimble provides both for an ICAR prior distribution and a proper PCAR distribution. I will demonstrate the ICAR here. The BYM model has two component random effects : a UH term and a CH term. The UH term has a zero mean Gaussian prior distribution and the CH term has a spatially correlated ICAR distribution. The BUGS code for the car.normal distribution can be converted to the nimble equivalent straightforwardly. The dcar_normal is the nimble equivalent to the car.normal in BUGS. Code for the convolution model is given below:

```
CONVmodel<-nimbleCode({
for(i in 1:m){
```

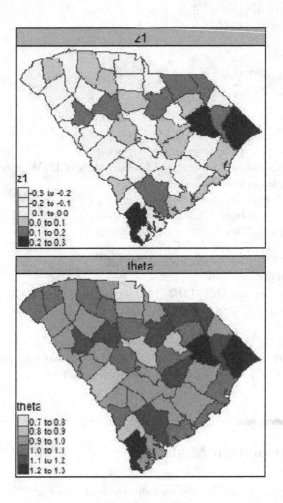

FIGURE 8.2
Posterior mean estimates of the uncorrelated effect ($z1$) and the relative risk
(θ_i) for the log normal model for the SC respiratory cancers example, using
nimble with 9000 iteration burnin and 1000 sample size.

```
Y1998[i]~dpois(mu[i])
mu[i]<-Exp98[i]*theta[i]
log(theta[i])<-a0+v[i]+u[i]
v[i]~dnorm(0,tauV)
}
u[1:m]~dcar_normal(adj[1:L],wei[1:L],num[1:m],tauU,zero_mean=1)
for(k in 1:L) {wei[k] <- 1 }
a0~dnorm(0,tau0)
tauV~dgamma(2,0.5)
tau0~dgamma(2,0.5)
tauU~dgamma(2,0.5)
})
```

Note that nimble requires that vectors in models have their dimensions specified and so the call to dcar_normal has dimensions included. In addition, there is a zero_mean parameter that should be set to equal 1 to ensure that the CAR component is zero centered. The data is entered as for the other model but the constant list now includes the adj and num vectors and the sum of number of neighbors (L). The WAIC for this model is 317.65. This is slightly higher than the LN model, but lower than the GP model. Figure 8.3 displays the UH, CH, and theta posterior mean estimates for this model.

It is often the case that an overall measure of goodness of fit is used to monitor convergence. The deviance can always be computed for a given data model by computing the item-wise log-likelihood ($-2 \log L_i$). For the Poisson model it is

```
LDev[i]<-2*(Y1998[i]*log(mu[i]+0.001)-(mu[i]+0.001)-lfactorial(Y1998[i]))
```

This can be summed to give a deviance and monitored via traceplot or other diagnostics.

```
geweke.plot(as.mcmc(samples$samples[,"Dev"]),main="Deviance")
traceplot(as.mcmc(samples$samples[,"Dev"]),main="deviance")
```

Figures 8.4 and 8.5 display the traceplot and Geweke diagnostic plot from CODA for the deviance (Dev) of the convolution model. The mean deviance is estimated as 295.012.

Both plots appear to suggest convergence in that the traceplot is relatively stable and centered while Geweke displays the range of the sample deviance to be within -2 and +2.

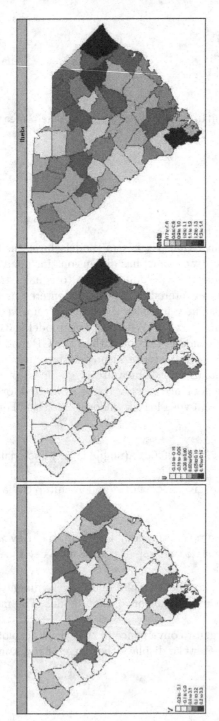

FIGURE 8.3
Posterior mean estimates for the UH , CH, and relative risk (θ_i) parameters for the convolution model applied to the respiratory cancers example.

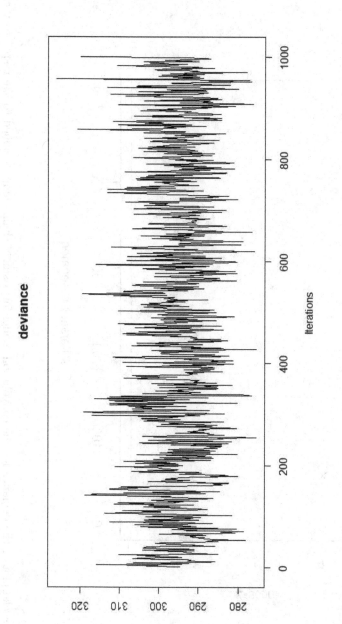

FIGURE 8.4
CODA traceplot of deviance for the convolution model for single chain sample applied to the respiratory cancers example

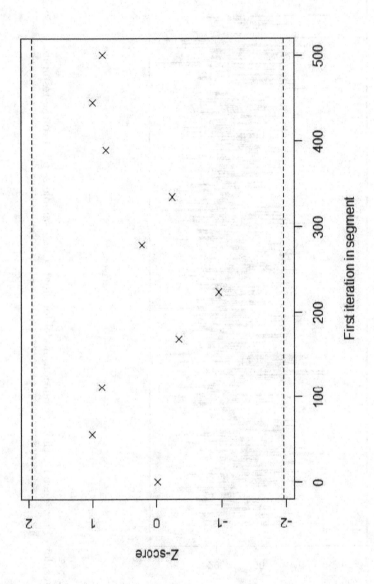

FIGURE 8.5
CODA Geweke plot of the deviance for the convolution model for a single chain sample applied to the respiratory cancers example.

8.4   Simple Mixture Model

A simple mixture model can also be fitted within nimble as follows:

```
SMmodel<-nimbleCode({
 for(i in 1:m){
 Y1998[i]~dpois(mu[i])
 mu[i]<-Exp98[i]*theta[i]
 log(theta[i])<-a0+p[i]*v[i]+(1-p[i])*u[i]
 v[i]~dnorm(0,tauV)
 p[i]~dbeta(1,1)
 }
 u[1:m]~dcar_normal(adj[1:L],wei[1:L],num[1:m],tauU,zero_mean=1)
 for(k in 1:L) {wei[k] <- 1 }
 a0~dnorm(0,tau0)
 tauV~dgamma(2,0.5)
 tau0~dgamma(2,0.5)
 tauU~dgamma(2,0.5)
 })
```

This model assumes a non-informative prior distribution for the mixing parameter. More sophisticated prior distributions could be imagined. Although an alternative logit(p) specification could lead to predictor models for clustering. The WAIC for this model is 312.58 which is lower than the convolution or GP model but close to the log normal model. Figure   8.6   displays the posterior mean estimates for a single chain sample of 1000, with burnin 9000 iterations, for the mixing probability (p), the UH and CH terms, and relative risk (θ_i).

The P map suggest that the UH term is favored in the southern areas whereas the CH term is predominant in the north east and upstate. The WAIC being lower than the convolution model suggest that a constant convolution of UH and CH terms is not favored in this example.

8.5   Leroux Model

The Leroux model as programmed in BUGS is not currently available on nimble as it involved a cyclic graph. However the model can be programmed directly by using the full multivariate normal specification:

$$s \sim \mathbf{MVN}(\mathbf{0}, \tau^2 Q(W, \rho)^{-1})$$

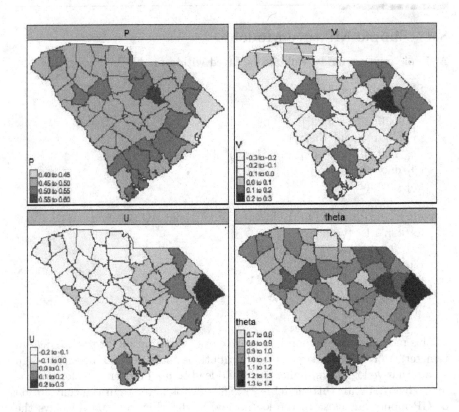

FIGURE 8.6
Posterior mean estimate maps for the mixing probability, UH, CH, and relative risk parameters for the simple mixing model applied to the respiratory cancers example.

where $Q(W, \rho)$ is a precision matrix specified by $\rho[diag(W1) - W] + (1 - \rho)I$ where W is a binary $m \times m$ neighborhood matrix. Here a full multivariate normal specification is used in nimble to estimate the spatial effects. To do this the precision matrix must be specified first. Note that the matrix is only stochastically dependent on ρ and so can be precomputed in R. The initial code is given below using the spatial polygon object SCmap (read in from shapefile using library sf). The binary weight matrix is created as W.mat and Q1 is the final precision matrix.

```
SCpoly<-st_read("SC_county_alphasort.shp")
SCmap<-as_Spatial(SCpoly)
```

```
W.nb <- poly2nb(SCmap)
W.mat <- nb2mat(W.nb, style="B")
adjnum<-nb2WB(W.nb)
num<-adjnum$num
### W.mat is a binary neighbor matrix #######
RD<-matrix(0,nrow=46,ncol=46)
R<-rep(1,46)
RD<-diag(R)
R0<-rep(0,46)
diaN<-diag(num[])
Q1<-diaN-W.mat-RD
```

The model code is then given below:

```
LERmodel<-nimbleCode(
{for (i in 1:m){Y1998[i]~dpois(mu[i])
p.sig[i]<-step(s[i])
log(mu[i])<-log(Exp98[i])+log(theta[i])
log(theta[i])<-a0+s[i]}
QRho[1:m,1:m]<-prec*(rho*Q1[1:m,1:m]+RD[1:m,1:m])
s[1:m]~dmnorm(R0[1:m],QRho[1:m,1:m])
a0~dnorm(0,tau0)
tau0~dgamma(2,0.5)
rho~dunif(0,1)
prec~dgamma(1,0.5)
})
```

The parameter rho (ρ) specifies the degree of spatial dependence in the model and is given a uniform prior distribution $U(0,1)$. The posterior mean estimate of ρ is 0.698 (SD: 0.181) with 95% credible interval estimated as (0.302,0.972). This is close to the BUGS estimate, based on the conditional code, for the same example (0.760). This suggests that there is strong evidence that overall spatial dependence is favored for a single effect under the Leroux model. The posterior mean estimates of the spatial effect and relative risk are displayed in Figure 8.7. The WAIC for this model is 308.03, which is lower than the convolution and LN and GP models.

8.6 BYM2 model

The final model considered here is the BYM2 model. The nimble version of this model follows closely to the BUGS version and also utilizes the INLA

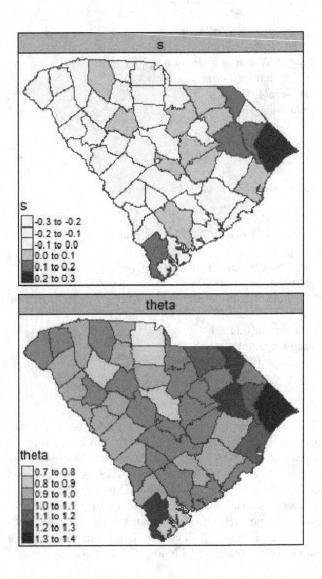

FIGURE 8.7

Posterior mean estimates of the spatial effect (s) and the relative risk (θ_i) for the Leroux model using nimble based on a single chain sample of 1000, after burnin of 9000 iterations, applied to the respiratory cancers example.

function inla.scale.model. In the code below the INLA graph file SC_poly.txt is used. This is used to generate a scaling of the SC counties based on the topology of the regions. The code also uses the functions inla.read.graph and inla.graph2matrix.

```
library(INLA)
#SC_poly.txt is the graph file
g=inla.read.graph("SC_poly.txt")
Q=-inla.graph2matrix(g)
diag(Q) = 0
diag(Q) = -rowSums(Q)
n=dim(Q)[1] # n=46
Q.scaled=inla.scale.model(Q,constr=list(A=matrix(1,1,n),e=0))
scale=Q.scaled[1,1]/Q[1,1]
```

The scale of the SC county graph is 0.3783966. The value is directly entered into the code for the BYM2 model. The code is as follows:

```
BYM2<-nimbleCode({
for(i in 1: n){
Y1998[i]~dpois(mu[i])
log(mu[i])<-log(Exp98[i])+log(theta[i])
theta[i]<-exp(ac0+bym2[i])
vc[i]~dnorm(0,1)
Corr[i]<-sigma*uc[i]*sqrt(rho/scale)
UCorr[i]<-sigma*vc[i]*sqrt((1-rho))
bym2[i]<-Corr[i]+UCorr[i]
}
for (k in 1 :nsumN){wei[k]<-1}
uc[1:n]~dcar_normal(adj[1:nsumN],wei[1:nsumN],num[1:n],1,zero_mean=1)

sigma~dunif(0,10)
tau<-pow(sigma,-2)
rho~dbeta(1,1)
ac0~dnorm(0,tau0)
tau0<-pow(sd0,-2)
sd0~dunif(0,10)
})
```

Figure 8.8 displays the posterior mean estimates for the uncorrelated (Ucorr) and correlated (Corr) random effects, and the relative risk for the BYM2 model fitted to the respiratory cancers example. Note that the ranges of the two components are relatively close, although the uncorrelated appears to

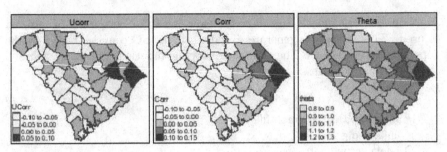

FIGURE 8.8
Posterior mean estimates of the uncorrelated, correlated components, and relative risk for the BYM2 model fitted to the SC county level respiratory cancers example, using a single chain sample of 1000, with burnin of 9000 iterations on nimble.

have positive areas correlated with the correlated component and does not seem to represent random noise as would be expected if the model fitted well.

The WAIC is estimated as 315.398 for the respiratory cancers data example. This is higher than the Leroux and other models, but marginally lower than the convolution model. The other parameters have the following estimates : $\rho : 0.5079\ (0.2636);$ and $\sigma : 0.1075\ (0.0282).$

9

CARBayes

CARBayes is a command based R package that fits spatial models to small area health data. To quote the package vignette for version 5.1.3:

"The package implements univariate and multivariate spatial generalised linear mixed models for areal unit data, with inference in a Bayesian setting using Markovchain Monte Carlo (MCMC) simulation. The response variable can be binomial, Gaussian, multinomial, Poisson or zero-inflated Poisson (ZIP), and spatial autocorrelation is modelled by a set of random effects that are assigned a conditional autoregressive (CAR) prior distribution. A number of different models are available for univariate spatial data, including models with no random effects as well as random effects modelled by different types of CAR prior. Additionally, a multivariate CAR (MCAR) model for multivariate spatial data is available, as is a two-level hierarchical model for modelling data relating to individuals within areas."

The package was first developed in 2012 and introduced via Lee (2013). It was extended with a spatio-temporal version called CARBayesST which is now in version 3.0.2, as of December 2019. The package has been developed with a command -driven format which is similar to the setup for commonly used linear and generalized linear model fitting on R. The package allows the fitting of a range of CAR-based models in a flexible linear models framework, using McMC with the fast Metropolis adjusted Langevin algorithm (MALA) as a default updater. The package allows single chains only and automatically generates output for the Geweke convergence diagnostics for single chains. The package also computes DIC, WAIC, and LMPL goodness of fit measures.

The steps for setting up a model fit are simple:

- define a formula

- fit model using formula

With these steps, the main output is the following: summary results as a table, a list of sampled values, fitted values and residuals, model fit, and acceptance probability (rate) for parameter sampling. The package exploits the common structure found in lm and glm model functions in R, in that a formula definition is followed by a one line model fit statement.

9.1   Models Available

Data models available on CARBayes are the binomial, Poisson, Gaussian, and zero-inflated Poisson (ZIP). These are treated as special cases within a generalized linear spatial model with a link function to a linear or non-linear predictor. The predictor can contain combinations of predictor or factor terms. The canonical links for these models are used: logit, log, identity, and log for ZI Poisson term.

The basic types of spatial models that can be defined are:

S.glm: generalized linear model with no random effects

S.CARbym: this fits a convolution model with a ICAR spatially structured effect

S.CARleroux: this fits the Leroux spatially structured effect

S.CARdissimilarity: this allows stochastically varying boundary definitions for adjacencies

S.CARlocalised: this fits a localized autocorrelation model using partitions

The basic GLM model has full call as:

```
S.glm(formula, formula.omega=NULL, family, data=NULL, trials=NULL,
burnin, n.sample,
    thin=1, prior.mean.beta=NULL, prior.var.beta=NULL, prior.nu2=NULL,
    prior.mean.delta=NULL, prior.var.delta=NULL, verbose=TRUE)
```

For a simple Poisson regression with two predictors x1 and x2 (say), based on simulated data:

```
x1<-rnorm(100,0,3);x2<-rgamma(100,2,1)
y<-rpois(100,exp(x1+x2))
```

The simple call could consists of:

```
form1<-y~x1+x2
res1<-S.glm(form1,family="poisson",burnin=20000,n.sample= 24000)
```

This produces the output
res1$summary

The full model fit details are available in res1 itself, including goodness of fit measures DIC and LMPL. Note that the median rather than the mean is produced along with 95% credible interval, sample size, acceptance rate, effective sample size, and a single chain is run. Convergence of parameters

	Median	2.50%	97.50%	n.sample	%	accept	n.effective	Geweke.diag
(Intercept	0.0394	-0.0197	0.0995	4000		44.3	27.3	-0.4
x1	0.9939	0.9866	1.001	4000		44.3	23.2	0.3
x2	1.0019	0.9912	1.0118	4000		44.3	67.6	0.6

FIGURE 9.1
Summary results from a CARBayes model fit.

is measured using the Geweke diagnostic. In this example all the Geweke Z scores are within the range +1.96 and -1.96 and so it could be concluded that the model has converged after a burnin of 20000 iterations.

Some output variants are also available. For example, for goodness of fit the command res1$modelfit will yield DIC, LMPL, and WAIC with effective numbers of parameters. For the object res1 we have;

res1$summary.results   a table of parameters estimates and convergence diagnostics

res1$samples   a list object containing the parameter samples and fitted values

res1$fitted.values   a list of fitted values for the data points

res1$residuals   a list of responses and Pearson residuals

res1$modelfit   a list of goodness of fit measures: DIC, pD, WAIC, pW, LPML, and loglikelihood.

9.2   Spatial Models

Fitting spatial models with CARBayes follows on the same lines as for the simple regression models with a few extra steps. The kernel of CARBayes is the Leroux spatial model and this model is used throughout. As special cases it is possible to fit a log normal model (Leroux with $\rho = 0$), a ICAR model (Leroux with $\rho = 1$) and a full Leroux model where ρ is estimated. The S.CARleroux function is used for this. It is possible to fit a convolution model but a different function must be used: S.CARbym.

9.3   Setting Up Spatial Information

For CARBayes, spatial information must be in the form of a binary weight matrix. This matrix is square with rows and columns representing regions.

Entries consist of 1 when two regions are neighbors and 0 otherwise. This can be set up from spatial polygon objects as follows using various libraries in R. For input from shapefiles use library(sf):

```
library(sf)
SCpoly<-st_read("...shapefile.......shp")
SCmap<-as_Spatial(SCpoly)
```

To get a polygon object from a Splus export then readSplus from library maptools must be used:

```
library(maptools)
SCmap<-readSplus("...........export file.txt")
```

Polygon objects can be converted to weight matrices using two functions from library spdep:

```
library(spdep)
W.nb <- poly2nb(SCmap)
W.mat <- nb2mat(W.nb, style="B")
```

W.nb is a neighborhood object that can also be interrogated to give the adj and num vectors used in nimble and BUGS.

```
adjnum<-nb2WB(W.nb)
num<-adjnum$num
adj<-adjnum$adj
```

W.mat is in the binary weight format that CARBayes uses.

9.4   Log Normal Model

For the log normal UH model the S.CARleroux function is used. First, for the basic random effect Poisson model without predictors an offset term must be used to allow for the expected counts in different areas. To do this the offset function can be used in the formula statement

```
form<-Y1998~1+offset(log(Exp98))
modelUH<-S.CARleroux(form,family="poisson",data=SCresp, W=W.mat,
rho=0, burnin=10000, n.sample=11000)
```

```
>modelUH$summary
            Median   2.5%   97.5% n.sample % accept n.effective Geweke.diag
(Intercept) 0.0018 -0.0353 0.0486    1000     39.8      199.7       -0.3
tau2        0.0093  0.0037 0.0219    1000    100.0       44.1       -1.6
rho         0.0000  0.0000 0.0000     NA       NA         NA         NA
>modelUH$modelfit
  DIC          p.d       WAIC        p.w        LMPL        loglikelihood

317.09041 16.42516 320.19698 14.70738  -162.87209      -142.12005
```

This model has DIC=317.09 and WAIC= 320.197. The prior distribution for the precision parameter (τ^{-2}) is defaulted to *gamma*$(1, 0.01)$, whereas in the model the variance is estimated (τ^2). The default prior for the intercept is zero mean Gaussian with large fixed variance of 100000.

MSPE is a commonly used predictive loss measure and this can be computed in CARBayes using posterior mean fitted values. The following code can be used to generate a predictive residual and then an MSPE.

```
sample.fitted<-modelUH$samples$fitted
m=46;L=1000
ypred<-matrix(0,ncol=m,nrow=L);ppl<-matrix(0,ncol=m,nrow=L)
for (j in 1:m){
for (i in 1:L){
ypred[i,j]<-rpois(1,sample.fitted[i,j])}}
ycopy<-matrix(0,ncol=m,nrow=L)
for(i in 1:m){
for(j in 1:L){
ycopy[j,i]<-SCresp$Y1998[i]}}
for(i in 1:m){
for( j in 1:L){
a<-(ycopy[j,i]-ypred[j,i])**2
ppl[j,i]<-a}}
mspel<-rep(0,m)
mspel<-colMeans(ppl)
mspe<-mean(mspel)
mspe
```

Figures 9.2 and 9.3 display the relative risk and UH random effect for the log normal model.

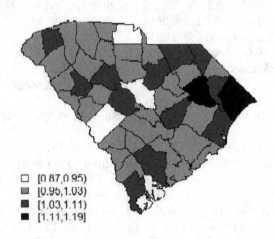

FIGURE 9.2
SC respiratory cancers: CARBayes relative risk estimates from the uncorrelated heterogeneity model using the S.CARleroux model with $\rho = 0$

9.5   ICAR Model

It is straight forward to change a model in CARBayes and so with the same S.CARleroux function an intrinsic CAR model can be fitted by setting $\rho = 1$. In this case the same formula can be used and the call is:

```
form<-Y1998~1+offset(log(Exp98))
modelCH<-S.CARleroux(form,family=" poisson",data=SCresp, W=W.mat,
rho=1, burnin=10000,
   n.sample=11000)
```

The overall model fit measures are DIC: 314.6161 pD: 11.735 WAIC: 316.51714 pW: 11.04445 LMPL: -159.23952. It would appear that the ICAR model fits slightly better than a simpler UH model as the DIC is slightly lower. Figure 9.4 displays the CH effect for this model.

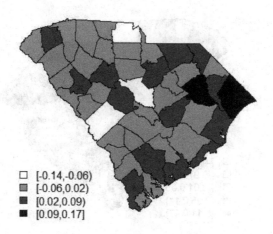

FIGURE 9.3
UH random effect for the CARBayes UH model for the SC respiratory cancers example.

9.6   Convolution Model

The BYM convolution model can be fitted on CARBayes via the **S.CARbym** function. For the respiratory cancers data the model is

```
form<-Y1998~1+offset(log(Exp98))
modelCONV<-S.CARbym(form,family="poisson",data=SCresp, W=W.mat,
burnin=10000,
  n.sample=11000)
```

The fit for this model is DIC: 315.10206 pD: 15.85755 WAIC: 315.40362 pW: 12.56809 and LMPL: -160.58557. Figure 9.5 displays the map of the estimated posterior median relative risk (θ_i) for a sample size of 1000, with burnin of 10000 iterations.

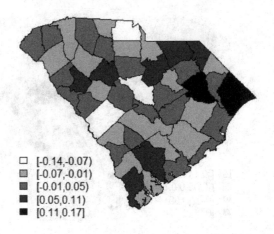

FIGURE 9.4
SC respiratory cancers posterior median CH effect from CARBayes ICAR model

9.7 Leroux Model

As noted above the Leroux model can be fitted where the ρ parameter is allowed to be estimated. It is assumed that it has a uniform $U(0,1)$ prior distribution. The code for this model is :

```
form<-Y1998~1+offset(log(Exp98))
modelLER<-S.CARleroux(form,family="poisson",data=SCresp, W=W.mat,
burnin=50000, n.sample=55000)
```

Convergence was not achieved at 10000 iterations with this model. The posterior median estimate, for a sample of 5000, for ρ is 0.333 with wide 95% credible interval (0.012, 0.889). Figure 9.6 displays the posterior median relative risk under the full Leroux model for the SC respiratory cancers example.

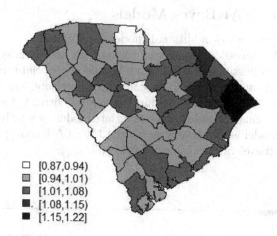

FIGURE 9.5
SC respiratory cancers example: median relative risk estimate map from the convoluti

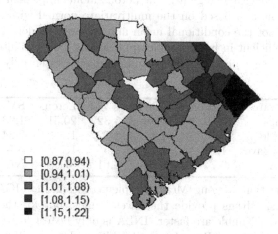

FIGURE 9.6
SC respiratory cancers posterior median relative risks map for the Leroux model

9.8   Other CARBayes Models

There are a number of additional models available on CARBayes. For example, a ZIP regression model can be used for zero-inflated data. ZIP regression is also possible on BUGS and Nimble via special programming. The S.CARdissimilarity and the S.CARlocalised, respectively, are useful when you want to model discontinuities within the spatial correlated surface, and also localized clustering. In addition to univariate models, a multivariate version of the Leroux model is now available via the MVS.CARleroux() function. There is also a multilevel variant using S.CARmultilevel().

9.9   Timing of CARBayes, Nimble, and BUGS models

To assess the computational requirements of the different packages, I report here a comparison of the timings for the different models, UH, ICAR, BYM, BYM2, SMIX, and the Leroux models for the different McMC softwares and INLA. I used system.time to assess the demand and assumed a burnin of 10000 with sample of subsequent 1000. This does not imply convergence but simply a benchmark for comparison. Timing is in seconds for fitting the SC county level cancer example with 46 regions. The Leroux model implementation for BRugs and Nimble here is based on the multivariate normal with scaled precision matrix, and not the conditional mean and variance model. This formulation will not be efficient in large sample applications, but it is interesting to note the time difference between BRugs and Nimble for essentially the same coded model.

Software	UH only	ICAR	BYM	Leroux	SMIX	BYM2
BRugs	0.79	0.89	1.32	20.51	1.98	5.49
Nimble	0.44	1.1	1.52	1.7	2.24	1.83
CARBayes	1.2	1.3	1.9	1.7	NA	NA
INLA	0.74	0.84	1.61	37.61	NA	12.80

Its notable that, among McMC implementations, for ICAR, SMIX, and BYM models, BRugs provide the lowest times, but for the Leroux model CARBayes and Nimble are faster. INLA is only fastest for the ICAR model. Nimble is faster than BRugs for the BYM2 model also. Note that the Nimble implementation does not exploit any sparse matrix processing, and is essentially coded as per the BRugs model both for Leroux and BYM2. In comparison of Nimble and CARBayes it seems that for most models Nimble is faster. These results don't reflect performance in general and are confined to the dataset and run time assumed. Convergence is not necessarily found for a burnin of 10000 iterations for McMC sampling. Convergence can vary between models and samplers used. It is notable that for these spatial models INLA does not provide the fastest computation overall.

10

INLA and R-INLA

A development in the use of approximations to Bayesian models has been proposed in a sequence of papers by Rue and coworkers (Rue et al., 2009; Lindgren et al., 2011; Simpson et al., 2012; Blangiardo et al., 2013; Lindgren and Rue, 2015; Rue et al., 2016). The basic idea is that a wide range of models which have a latent Gaussian structure can be approximated via integrated nested Laplace approximation (INLA). These approximations can be seen as successive approximations of functions within integrals. The integrals are then approximated by fixed integration schemes and not using Monte Carlo integration. If we consider a Poisson model for observed counts: y_i, $i = 1, ..., m$ then with the set of hyperparameters given by ϕ and a log link to an additive set of effects (random effects) then

$$y_i|\boldsymbol{\lambda}_i \sim Pois(e_i\theta_i)$$
$$\theta_i = \exp\{\alpha + v_i + u_i\}$$
$$\text{where } \boldsymbol{\lambda}_i = \{\alpha, v_i, u_i\}^T.$$

Note that the parameters in $\boldsymbol{\lambda}$ all have Gaussian distributions and have prior distribution $P(\boldsymbol{\lambda}|\phi)$. In this case it is possible to approximate the posterior marginal distribution $P(\boldsymbol{\lambda}_i|\boldsymbol{y})$ by

$$P(\boldsymbol{\lambda}_i|\,\boldsymbol{y}) = \int_{\phi} P(\boldsymbol{\lambda}_i|\boldsymbol{y}, \phi)P(\phi \mid \boldsymbol{y})d\phi.$$

for each component λ_i of the latent fields. The terms $P(\boldsymbol{\lambda}_i|\boldsymbol{y}, \phi)$ and $\boldsymbol{P}(\phi|\boldsymbol{y})$ can each be approximated by Laplace approximation. The simplest of these is the Gaussian approximation where matching of the mode and curvature to a normal distribution is used. Finally the integral approximation leads to

$$\widetilde{P}(\boldsymbol{\lambda}_i|\,\boldsymbol{y}) = \sum_k \triangle_k \widetilde{P}(\boldsymbol{\lambda}_i|\boldsymbol{y}, \phi_k)\widetilde{P}(\phi_k|\boldsymbol{y}).$$

This approximation approach is now available in R (package R-inla: www.r-inla.org), and can be used for a wide variety of applications. Application of these approximations has been made by Schrodle et al. (2011) to veterinary spatial surveillance data. In that work they demonstrate the closeness of the final estimates to that achieved using posterior sampling within McMC. However they do not show any simulated comparisons where a ground truth is compared. Further demonstration of the capabilities in spatio-temporal modeling is given by Schrodle and Held (2011). A recent simulated comparison of the performance of INLA to McMC for disease mapping can be found in Carroll et al. (2015).

In a recent extension, the parallel with finite element solutions to differential equations is exploited by Lindgren et al. (2011) whereby the spatial field is a solution to a stochastic partial differential equation (SPDE) with form $\lambda(s_i) = \sum_k \phi_k(s)w_k$ where the $\phi_k(s)$ are basis functions and w_k are weights. This is formally close in form to the kernel process convolution models of Higdon (2002). A comparison is made by Simpson et al. (2012). Some examples of the use of INLA for Bayesian disease mapping examples is given in Lawson (2018), chapter 15. A link to a tutorial on INLA can be found here: https://www.precision-analytics.ca/blog-1/inla. A recent text describing the use of INLA for Bayesian modeling is Gomez-Rubio (2020).

Here I will focus on the use of INLA for the basic spatial models described in previous chapters. As in the case of CARBayes, INLA is based on definition of a formula and model fit statements. The variables used in the model fitting must reside in a dataframe. The basic form of a simple linear regression with Gaussian error is similar:

```
x1<-c(1.1,2.3,3.4,4.5,5.4)
x2<-c(-2.3,4.5,3.6,6.8,12.7)
y<-c(1.2,1.4,2.3,3.2,1.2)
As<-data.frame(x1,x2,y)
formula1<-y~1+x1
res1<-inla(formula1,family="gaussian",data=As,control.compute=list(dic=
TRUE,cpo=TRUE))
summary(res1)
```

In this model fit a simple Gaussian regression model is fitted with data in dataframe As. The control.compute=list() statement lists what should be computed as well. In this case the DIC and CPOs are specified.

The summary of the fit summary(res1) includes the model fitted, posterior mean parameter estimates, and summary of goodness of fit measures (DIC, marginal likelihood). The fitted object res1 includes a wide range of additional summary results including: summary.fixed, summary.random, summary.fitted.values, summary.linear.predictor. Included in the components are also the DIC (local DIC and pD), CPO (already local) and the WAIC (if these are specified in control.compute). Note that a number of default specifications are assumed here. First the model is

$$y_i \sim N(\mu_i, \tau_y^{-1})$$
$$\mu_i = \alpha_0 + \alpha_1 x1$$

and the following are default assumptions made by INLA:

$\alpha_0 \sim U(-10000, 10000)$

$\alpha_1 \sim N(0, \tau_1^{-1})$

$\tau_y \sim gamma(1, 0.00005)$

$\tau_0 \sim gamma(1, 0.00005)$

$\tau_1 \sim gamma(1, 0.00005)$.

It is possible to change these default specifications to a certain extent. For example, the parameters of the gamma prior distributions can be specified via a call to : control.family(). However prior distributions themselves can only be changed by specifying a special prior distribution function for each parameter. Random effects can be added to these models straightforwardly using the f() function. For example for an uncorrelated Gaussian random effect added to the simple regression above we could have code:

```
ind=seq(1:m)
formula2<-y~1+x1+f(ind,model="iid",param=c(2,0.5))
res2<-inla(formula2,family="gaussian",data=As,control.compute=list(dic=TRUE,cpo=TRUE))
```

To fit such effects an index vector must be set up which identified the unit to which a random quantity is assigned. In this case ind is just assigned the sequence 1:m, where m is the number of items. In this formulation the gamma prior distribution for the precision of the random effect has parameters (2,0.5) specified in param=c(,).

10.1 Uncorrelated Heterogeneity (UH models)

UH models can be fitted straightforwardly on INLA. All that is needed is the index vector assigned to each spatial unit. For the SC respiratory cancers example this would be m=46: region=seq(1:m). Then the model formula and fit statements are

```
formulaUH = Y1998~f(region, model = "iid",param=c(2,0.5))
resUH = inla(formulaUH,family="poisson",data=SCresp,control.compute=list(dic=TRUE,cpo=TRUE,waic=TRUE),E=Exp98)
```

This model fits a random intercept model under a Poisson data model with log(Exp98) as offset. resUH returns the precision estimates for the random effect and intercept terms. The posterior mean effect estimates can be obtained by interrogating the resUH$summary.random$region. This is a dataframe containing the summary measures on the parameter vector region. The posterior mean can be obtained from resUH$summary.random$region["mean"], the SD

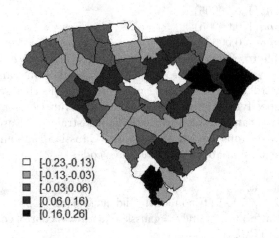

FIGURE 10.1
Posterior mean map of UH effect for the log normal model using INLA for the SC respiratory cancers example.

from resUH$summary.random$region["sd"] and also other quantiles. Figure 10.1 displays the UH effect posterior mean map for this model.

The CPO is locally estimated and it can also be mapped as can the local DIC or WAIC.
The CPO is:

```
cpo<-resUH$cpo$cpo
```

The local DIC is:

```
locdic<-resUH$dic$local.dic
```

The LMPL can also be computed from the sum of the logged cpos:

```
LMPL=sum(log(cpo)).
```

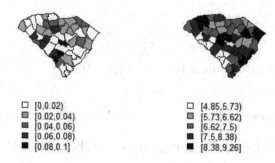

□ [0,0.02)	□ [4.85,5.73)
▣ [0.02,0.04)	▣ [5.73,6.62)
▩ [0.04,0.06)	▩ [6.62,7.5)
■ [0.06,0.08)	■ [7.5,8.38)
■ [0.08,0.1]	■ [8.38,9.26]

FIGURE 10.2
CPO (left) and local DIC (right) maps for the log normal model fitted on INLA for the SC county level respiratory cancer example.

Figure 10.2 displays the CPO and local DIC maps for these data. Notice that the CPO ranges between 0 and 1 with values close to 1 indicating a good fit. The local DIC shows good fit where the DIC is lowest. Hence the two maps can be regarded inverse of each other. In this case the model fits well in the W/SW part of the state.

For this model fit the following overall diagnostics are found : DIC= 323.77, pD= 29.88, WAIC=312.97, pW= 14.36, and LMPL= -166.3929. Note that if you wanted to obtain an MSPE it is more difficult to do that with INLA, as the predictive distribution is not directly available. An approximate version using the fitted model values can be obtained using:

```
ypred<-matrix(0,ncol=1000,nrow=m);ppl<-matrix(0,ncol=1000,nrow=m)
for(i in 1:1000){
for(j in 1:m){
ypred[j,i]<-rpois(1,resUH$summary.fitted.values[j,1])
}}
ycopy<-matrix(0,ncol=1000,nrow=m)
for(i in 1:1000){
for(j in 1:m){
ycopy[j,i]<-SCresp$Y1998[j]}}
for(i in 1:1000){
for( j in 1:m){
a<-(ycopy[j,i]-ypred[j,i])**2
ppl[j,i]<-a}}
mspel<-rep(0,m)
mspel<-rowMeans(ppl)
```

```
mspe<-mean(mspel)
mspe
```

For this example and model fit: MSPE=8416.17.

10.2 Correlated Heterogeneity (CH) models

Spatial structure can be modeled in INLA straightforwardly. The package allows the specification of ICAR and convolution models. INLA requires the use of a special graph file that includes the adjacencies of the study regions. An INLA graph file can be created from adj and num vectors as used by Bugs or Nimble using

```
inla.geobugs2inla(adj, num, graph.file=".........txt")
```

This file can also be created using function nb2inla from the spdep package. I will use a graph file already created: SCgraph.txt. This file has a head which looks like this:

```
46
1 5 33 30 24 23 4
2 5 41 38 32 19 6
3 4 25 15 6 5
4 5 39 37 30 23 1
5 5 38 25 15 6 3
6 4 38 5 3 2
7 3 27 25 15
.
.
.
.
```

The graph file has the following structure. The first element is the number of regions (in this case 46). The first column is an index to the regions and the second column is the number of neighbors of that region. The rest of the row is the labels of the regions that are neighbors. For example region 1 has 5 neighbors and they are regions 33, 30, 24, 23, and 4. For CH models this graph file must be specified in the model formula.

10.2.1 ICAR model

The ICAR model can be fitted using the "besag" model specification. The specification is

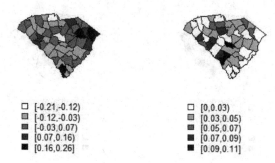

□ [-0.21,-0.12)
▣ [-0.12,-0.03)
■ [-0.03,0.07)
■ [0.07,0.16)
■ [0.16,0.26]

□ [0,0.03)
▣ [0.03,0.05)
■ [0.05,0.07)
■ [0.07,0.09)
■ [0.09,0.11]

FIGURE 10.3
Posterior mean ICAR effect (left) and cpo map (right) for the ICAR model
fitted to the SC respiratory cancers data using INLA.

```
region<-seq(1:m)
  formula<-Y1998~1+f(region,model="besag",param=c(2,0.5),graph=
"SCgraph.txt")
```

This specifies the ICAR model for index vector region with a gamma prior
distribution for the precision with parameters (2,0.5). The intercept is assumed to have a default large range uniform prior distribution ($U(-a,a)$,
where a is large). Figure 10.3 displays the posterior mean ICAR effect (left)
and the CPO map for this model fit to the SC respiratory cancers example.
The CPO map suggests a better fit in west and south west in this case.

The goodness of fit metrics for this model are DIC= 318.46, pD=21.86;
WAIC= 313.94;pW=13.35 ; LMPL=-162.94.

10.2.2 The convolution model

The convolution or BYM model can be fitted in two different ways in INLA.
First a two component model can be specified with a zero mean Gaussian
random intercept term and an ICAR term. Second there is a composite model
which fits both terms. The latter is specified by model="bym" in a single f(.)
specification. The example given here is for the first approach where two
components are fitted. The code for this is

```
region<-seq(1:m);region2<-region
  formCONV = Y1998~1+f(region, model = "iid",param=c(2,0.5))+f(region2,
model = "besag",param=c(2,0.5),graph="SCgraph.txt")
  resCONV = inla(formCONV,family="poisson",data=SCresp,control.compute=
list(dic=TRUE,cpo=TRUE,waic=TRUE),E=Exp98)
```

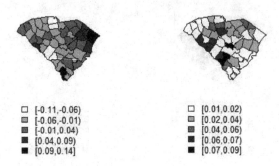

☐ [-0.11,-0.06)	☐ [0.01,0.02)
▥ [-0.06,-0.01)	▥ [0.02,0.04)
■ [-0.01,0.04)	■ [0.04,0.06)
■ [0.04,0.09)	■ [0.06,0.07)
■ [0.09,0.14]	■ [0.07,0.09]

FIGURE 10.4
Posterior mean map of the ICAR component (left) and the cpo map for the convolution model fitted on INLA for the SC respiratory cancers example.

In this model the precisions are assumed to have gamma (2,0.5) prior distributions. Figure 10.4 displays the posterior mean ICAR component and cpo map for the convolution model fitted to the SC respiratory cancers example. It is notable that the highest cpo areas are in the west and south west, whereas the positive clustering is markedly in the north east. The goodness of fit metrics for this model are: DIC=329.06, pD= 33.90; WAIC=316.61, pW=15.96, LMPL=-171.52 .

Of the basic spatial models fitted so far (log normal, ICAR, convolution) it appears that the ICAR model has the lowest DIC, and least negative LMPL, although similar WAIC to the log normal model for these data.

10.2.3   The Leroux model

The convolution model is not well identified (computationally) and so it could be useful to consider an alternative where the only one effect is estimated and the degree of spatial structure is estimated for the single spatial effect. The Leroux model is not available in the main INLA package but is available in INLABMA package. This basically iteratively supplies values of ρ to a leroux.inla function. The code relevant to our example is:

```
library(INLABMA)
library(spdep)
library(maptools)
W.nb <- poly2nb(SCpoly)
W.mat <- nb2mat(W.nb, style="B")
rlambda <- seq(0.03, 0.8, length.out = 20)
errorhyper <- list(prec = list(prior = "loggamma",
```

```
param = c(1, 0.01), initial = log(1), fixed = FALSE))
form2 <- Y1998 ~1+offset(log(Exp98))
lerouxmodels <- mclapply(rlambda, function(lambda) {
leroux.inla(form2, d =SCresp, W = W.mat,
lambda = lambda, improve = TRUE,
family = "poisson",
control.predictor = list(compute = TRUE),
control.compute = list(dic = TRUE, cpo = TRUE))
})
resLER<- INLABMA(lerouxmodels, rlambda, 0, impacts = FALSE)
```

Note that this code estimates the mixing parameter ρ with value 0 implying a ICAR model and 1 a log normal model. This is the reverse of the parameterization for Bugs and Nimble models. The INLA posterior estimate of ρ is 0.3917 (sd: 0.2252). This implies a stronger evidence of spatial structure in these data. The DIC for this model is 322.96 with pD = 11.99. The LMPL is -165.47. The WAIC is not immediately available however.

10.2.4 The BYM2 model

The BYM2 model was developed to compensate for the fact that the two component convolution model does not have easily identified components.

For the SC respiratory cancers example the posterior mean estimate of the ϕ is 0.33 (sd: 0.2610). This implies that the mean proportion of the marginal variance explained by the spatial effect is 0.33, after standardization of the generalized variances of the two components to have value 1.

```
formBYM2 = Y1998~1+f(region2, model = "bym2",scale.model=TRUE,
param=c(2,0.5),graph="SCgraph.txt")
resBYM2 = inla(formBYM2,family="poisson",data=SCresp,control.compute=
list(dic=TRUE,cpo=TRUE,waic=TRUE),E=Exp98)
```

DIC=316.00, pD=16.88; Waic=315.38, pW= 12.69, LMPL= -161.44. Figure 10.5 displays the UH and CH components for the SC cancers example. The parameter estimate suggests that there is not strong evidence of a spatial effect although there is clearly some identification issue in that the UH effect also seems to include some clustering in the north east.

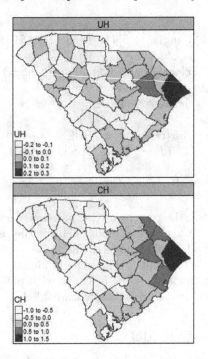

FIGURE 10.5
Posterior mean maps of the UH and CH effects for the BYM2 model fitted to
the SC respiratory cancers example.

11

Clustering, Latent Variable, and Mixture Modeling

11.1   Clustering in Spatial health data

Assessment of clustering is important in small area health studies. Both public health authorities and epidemiological researchers benefit from being able to assess the degree to which disease clusters. This could suggest localized areas of adverse disease risk, with potential for further study or intervention. There are two basic approaches to assessment of clustering in Bayesian modeling. The first is exceedance estimation, and the second is the use of models designed to find clusters or clustering.

11.1.1   Exceedance

It is possible to assess how "extreme" estimated parameters are by assessing the tail area of the posterior marginal distribution. This can be estimated from a posterior McMC sample via a simple coding. How often the sampled parameter exceeds (upper tail) or deeds (lower tail) a particular prespecified level (c) is usually of interest. For elevated relative risk it is common to use an estimate of $\Pr(\theta_i > c)$, where c is the threshold level. For relative risk $c = 1$ is commonly assumed, although for more extreme risks $c = 2$ or $c = 3$ could also be used. Note that changing c will lead to different probability criteria: e.g. $\Pr(\theta_i > 1) = 0.95$ would be interpreted differently from $\Pr(\theta_i > 2) = 0.95$, the latter being much more extreme. $\Pr(\theta_i > c)$ can be thought of as a probability statement about the 'significance' of a region's risk. Code for computing this exceedance from sampled output, of size G, exploits the approximation:

$$\Pr(\theta_i > c) \approx \sum_{g=1}^{G} I(\theta_i^g > c)/G$$

where θ_i^g is the g th sampled value of θ_i. For BUGS or Nimble this is:

```
prex[i]<-step(theta[i]-1)
```

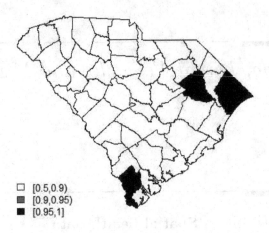

FIGURE 11.1
Exceedance probability of the relative risk mapped for $c = 1$. Convolution model using Nimble.

with **theta** estimated as the relative risk in the code then **prex** when averaged over the sample will yield an estimate of the probability of exceedance. In this case its $c = 1$. If a binomial model is assumed and exceedance of probability is required then the coding would be

```
prex[i]<-step(prob[i]-0.5)
```

where **prob** is the binomial probability. It should be noted however that exceedance probability estimation based on tail areas is highly dependent on the assumed model. If the model is misspecified then the exceedance could be markedly different from the correctly specified model. Lawson (2018) chapter 6, provides a dramatic example of this sensitivity (see figure 6.8). Some references to the original proposal for the use of exceedance for relative risk are Richardson et al. (2004), Abellan et al. (2008), and extension to neighborhoods with $L_i = \sum_{j=0}^{n_i} I(p_{ij} > 0.05)/(n_i + 1)$ where p_{ij} is the exceedance probability ($c = 1$) of neighbors of the i th region by Hossain and Lawson (2006). Figure 11.1 displays the exceedance probability estimate map for SC respiratory cancers example using a convolution model with threshold $c = 1$. Three counties have very high exceedance (>0.95): Horry, Florence, and Jasper counties.

Exceedance can be used to detect hot spot clusters where it is simply the focus to detect where regions have unusual risk. The method does not impose any restrictions on where the region lies in relation to other areas. Of course the L_i measure takes account of neighbors and their exceedances.

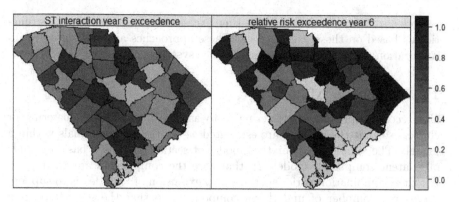

FIGURE 11.2
ST interaction maps for the interaction effect (left) and the relative risk (right) from the Knor-Held UHST model fitted on Nimble.

11.1.1.1 Space-time (ST) extensions

The exceedance probability concept can be extended to spatio-temporal data. Following Abellan et al. (2008), we assume that we model the relative risks θ_{ij} where j denoted the time period, for a Poisson data model, and we could fit a model such as $\log(\theta_{ij}) = \alpha_0 + v_i + u_i + \gamma_j + \psi_{ij}$. Note that both θ_{ij} and ψ_{ij} are indexed for each region and time period. We could examine overall exceedance by looking at $\Pr(\theta_{ij} > 1)$ for regions or time periods and this could be computed within the main code loop for a McMC sampler.

```
prex[i,j]<-step(theta[i,j]-1)
```

This would allow the examination of individual areas of elevated risk which are evolving over time. Cross sections in time can be examined also. More specifically it could be useful to examine $\Pr(\psi_{ij} > 0)$. This metric would allow spatio-temporal monitoring for the emergence of ST clusters i.e., spatial clusters which appear and disappear quickly within a short space of time. These may only be found in evaluation of ψ_{ij} surfaces. The code for this assuming psi[i,j] is the interaction term is

```
prexpsi[i,j]<-step(psi[i,j])
```

Figure 11.2 displays the mapped exceedances for year 6 from the fitted UHST model of Knorr-Held with interaction exceedance (left panel) and relative risk exceedance (right panel). It is notable that for some counties short term changes in risk are present, as exemplified by the left panel, whereas a mix of factors contributes to the risk profiles in the right panel.

Note also that cylinders of space-time excess risk can be estimated as functionals based on these exceedances. These approaches are well suited to incorporation within prospective surveillance systems.

11.1.2   Cluster Models

The second approach to clustering is to assume a model for clustering or clusters so that the clusters are estimated as entities or functionals within a model. The simple fixed mixture model of section 7.4.1 is a basic example of a latent component model. In that case the components are fixed (v, u) but their combination varies spatially. An extension of this idea is to suppose there are a number of underlying components so that the mean level is a combination of these.

For the classic Poisson risk model we could have

$$y_i \sim Pois(e_i\theta_i)$$
$$\log(\theta_i) = \alpha_0 + \sum_{l=1}^{L} w_{il}\phi_l. \qquad (11.1)$$

Here the ϕ_l, $l = 1 :, , ., L$ are components, and w_{il} are weights. These components represent unobserved groupings in the risk. These types of models have been proposed in different forms. For example, partition models such as those of Ferreira et al. (2002), or Knorr-Held and Rasser (2000) allow patches of risk to exist. In some cases models penalize overlaps. Grouping of components has been proposed by Lee and Sarran (2015) whereby a variant of the above mixture allows an assignment of areas to a group of G intercepts via a penalty loss and the $\log(\theta_i) = u_i + \phi_{z_i}$ where u_i is a spatial effect and the assigned ϕ_{z_i} is obtained from $f(z_i)$ a penalty and z_i is a label on the range $1, ..., G$. The ϕ_{z_i} are assumed to be ordered to avoid label switching. CARBayes has a function that fits this type of model: S.CARlocalised.

An example of the above model (11.1) can be found below in BUGS code

```
model{
  for( i in 1 : m ) {
  y[i] ~dpois(mu[i])
  ypred[i]~dpois(mu[i])
  mu[i] <- e[i] * rr[i]
      rr.temp[i] <- al0 + (w[i,1]*theta.s[1]) + (w[i,2]*theta.s[2]) + (w[i,3]
*theta.s[3]) +
    (w[i,4]*theta.s[4])   +   (w[i,5]*theta.s[5])   +   (w[i,6]*theta.s[6])   +
(w[i,7]*theta.s[7]) +
    (w[i,8]*theta.s[8]) + (w[i,9]*theta.s[9]) + (w[i,10]*theta.s[10])
    rr[i] <- exp(rr.temp[i])
```

```
mse.temp[i] <- pow((tht.true[i]-rr[i]),2)
mspe.temp[i]<- pow((y[i]-ypred[i]),2)
}
mse <- mean(mse.temp[])
mspe <- mean(mspe.temp[])
# weight prior with CAR prior
for (i in 1:m) {
for (l in 1:L) {
w[i,l] <- psi[l] * delta[i,l]/max(inprod(psi[],delta[i,]),exp(-30))
delta[i,l]<-exp(delta1[i,l])
delta1[i,l]~dnorm(mu.delta1[l,i], tau.delta1[l])
}
}
# For CAR prior
mu.delta1[1,1:m] ~car.normal(adj[], weights[], num[], tau.delta[1])
mu.delta1[2,1:m] ~car.normal(adj[], weights[], num[], tau.delta[2])
mu.delta1[3,1:m] ~car.normal(adj[], weights[], num[], tau.delta[3])
mu.delta1[4,1:m] ~car.normal(adj[], weights[], num[], tau.delta[4])
mu.delta1[5,1:m] ~car.normal(adj[], weights[], num[], tau.delta[5])
mu.delta1[6,1:m] ~car.normal(adj[], weights[], num[], tau.delta[6])
mu.delta1[7,1:m] ~car.normal(adj[], weights[], num[], tau.delta[7])
mu.delta1[8,1:m] ~car.normal(adj[], weights[], num[], tau.delta[8])
mu.delta1[9,1:m] ~car.normal(adj[], weights[], num[], tau.delta[9])
mu.delta1[10,1:m] ~car.normal(adj[], weights[], num[], tau.delta[10])
for(k in 1:sumNumNeigh) { weights[k]<-1 }
for(l in 1:L) {
tau.delta[l]<-pow(sigma.delta[l],-2)
sigma.delta[l] ~dunif(0,5)
tau.delta1[l]<-pow(sigma.delta1[l],-2)
sigma.delta1[l] ~dunif(0,5)
psi[l]~dbern(p.psi)
temp[l] ~dgamma(1,1)
}
#To ensure the inequality constraint
for (l in 1 : L){
    theta.s[l] <- sum(temp[1:l])
}
al0 ~dflat()
p.psi ~dbeta(1,1)}
```

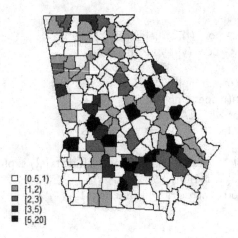

FIGURE 11.3
Standardized Incidence Ratio map of the Georgia oral cancer example.

This assumes that the weights are defined to have spatial structure via:

$$w_{il} = \psi_l . \delta_{i,l} / \max(\psi . \delta)$$
$$\delta_{i,l} = \exp(\delta_{1i,l})$$
$$\delta_{1i,l} \sim N(\mu_{\delta_1 l}, \tau_{\delta l})$$
$$\mu_{\delta_1 l} \sim \tau . \delta$$

and the $\theta.s_l$ are ordered so that they are identified.

Figure 11.3 displays the standardized incidence ratios for the Georgia counties oral cancer example. Figure 11.4 displays the posterior mean relative risk estimate map for the cluster model above

Figure 11.5 displays the posterior median maximum probability assignment for the cluster model for the Georgia data.

In this example, which allowed ten underlying components, the posterior median assignments based on max(weight) is for 3 components with the highest concentration on component 1 and 3, with a few areas on 2. This represents a predominantly low risk and high risk set. The posterior mean estimates of these components were $\theta_1 = 0.3561$ (0.381), $\theta_3 = 1.6556$ (0.747). The posterior weight estimates for the ten components is displayed in Figure 11.6

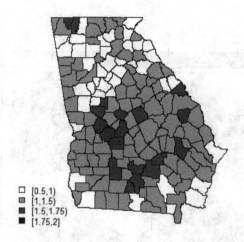

FIGURE 11.4

Georgia oral cancer example: posterior mean relative risk map under the cluster model

FIGURE 11.5

Posterior median max(weight) probability assignment to components under the cluster model.

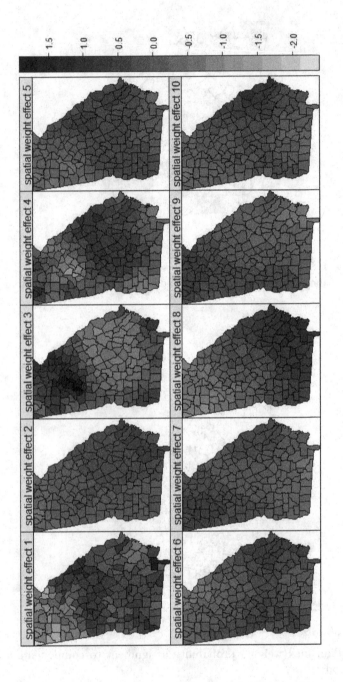

FIGURE 11.6
Spatial posterior weight estimates for 10 components

11.2 Dimension Reduction

Note that other types of latent models, exploiting dimension reduction approaches such as factor analysis or Principal components, are usually applied either to a multivariate vector of outcomes (Wang and Wall, 2003, Liu et al., 2005) in the spatial case, or to spatio-temporal realizations (Lopes et al., 2008). Specifically, applied to space-time data, the R package spBFA allows the fitting of spatio-temporal dynamic factor models for a range of outcome distributions (Berchuck et al., 2019).

In the spatial setting, it is often assumed that

$$y_{ik} \sim Pois(\mu_{ik})$$

where $k = 1, ..., p$ outcome variables and

$$\mu_{ik} = e_{ik}\rho_{ik} \tag{11.2}$$
$$\log(\rho_{ik}) = \lambda_k f_i$$

with λ_k defined to be the factor loading for the k th outcome and f_i underlying common factor. The common factor can have spatial structure and a proper CAR model is often assumed. An intercept is often not included as the offset expected count should allow for the different rates. The prior distributions are usually

$$\lambda_k \sim N(0, \tau_\lambda^{-1})$$
$$f_i | \{f_l\}_{l \neq i} \sim N(\phi\mu_{f_{\delta_i}}, \tau_f^{-1}/n_{\delta_i})$$
$$\mu_{f_{\delta_i}} = \sum_{j \in \delta_i} f_j/n_{\delta_i}$$
$$\phi \sim U(a, b)$$
$$\tau_* \sim gamma(c, d)$$

a, b are the inverse of the max and min eigenvalues of the weight matrix W. c and d are often weakly informative (e.g., $c = 2, d = 0.5$). The nimble code for this model is given below. To prepare for the proper CAR model nimble provides a function to convert adj and num vectors to matrices C and M (as.carCM) and also a function to compute the bounds for the spatial association parameter in the proper CAR model (carBounds):

```
CM2<-as.carCM(adj[1:L],weights[1:L],num[1:Nareas])
C<-CM2$C
M<-CM2$M
```

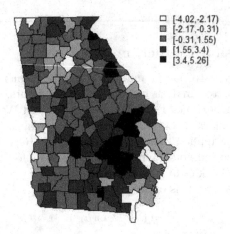

FIGURE 11.7
Posterior mean common factor f surface estimate: Georgia three disease example.

```
bou<-carBounds(C,adj,num,M)
if(bou[1]<0)bou[1]=0
FAmodel<-nimbleCode({
for(i in 1:Nareas){
for(k in 1:Ndiseases){
O[i,k]~dpois(mu[i,k])
mu[i,k]<-E[i,k]*rho[i,k]
log(rho[i,k])<-lambda[k]*fact[i]}}
phi~dunif(bou[1],bou[2])
fact[1:Nareas]~dcar_proper(mumod[1:Nareas],C[1:L],adj[1:L],num[1:Nareas],
M[1:Nareas],tauF,phi)
for (k in 1:Ndiseases){
lambda[k]~dnorm(0,tauL)}
tauL~dgamma(2,0.5)
tauF~dgamma(2,0.5)
})
```

For the three disease (asthma, COPD, Angina) Georgia county level example, the factor model gave the following converged posterior mean estimates: $\phi = 0.95469$ (sd: 0.03044), $\tau_f = 0.093255$ (sd: 0.03674), and $\tau_\lambda = 5.36969$ (sd: 2.91733), λ : (0.15491786, 0.31680317, 0.43252911), sd$_\lambda$: (0.03017, 0.06093, 0.08377). The posterior mean estimated f surface is provided in Figure 11.7.

The results of performing this dimension reduction are discussed later (in Chapter 14) when examining multivariate models in more depth.

12

Spatio-Temporal Modeling with MCMC

The extension of Bayesian small area health models to space-time varies in complexity depending on the nature of the observational process and the type of model envisaged. For a more extensive discussion of ST modeling see Lawson and Choi (2016), and Lawson (2018), ch. 12. Here I will focus on the simplest observational situation: where a map of small areas is repeatedly observed over discrete time periods. This is known as *map evolutions*. A common example could be the observation of counts of disease in provinces of a state over a period of months or years. Assuming the disease is recorded as new incident cases, then the following notation can be assumed: y_{ij} incident cases in i th region in j th time period, $i = 1, ..., m$ regions, and $j = 1, ..., T$ time periods. Also the expected count in the same unit is e_{ij} and the relative risk is θ_{ij} . Hence for a Poisson model we would assume $y_{ij} \sim Pois(\mu_{ij})$ and $\mu_{ij} = e_{ij}\theta_{ij}$. For finite population binomial models it could be assumed that the unit probability of disease is p_{ij} and the finite population is n_{ij} so that in this case the disease count has data model $y_{ij} \sim bin(p_{ij}, n_{ij})$. In both cases it is usual to have a link function that associates the parameter of interest with a linear or nonlinear predictor. For the Poisson case this is usually a log link while for binomial it is usually logit link. Other links are possible of course, the underlying criteria being that the parameter of interest is mapped to the infinite real line. Hence for a Poisson model $\log(\theta_{ij})$ is modeled while for a binomial model the $\text{logit}(p_{ij})$ would be the focus.

In what follows I will examine a small selection of the classic space-time models that have been proposed so far. These include the temporal trend model of Bernardinelli et al. (1995), the interaction models of Knorr-Held (2000) and a variant Kalman filtering approach. In addition, some simpler models are also examined with respect to goodness of fit for a given dataset. The space-time data set of county level SC all respiratory cancers for the years 2011 to 2016 will be examined (*SCRCST*)

12.1   Basic Models

I will focus here on Poisson data models. Most of the coding issues related to finite populations carry over directly. Separable models of spatial and temporal effects are the most basic type of model conceivable which includes both spatial (S) and temporal (T) components.

Models are of the form

$$y_{ij} \sim Pois(\mu_{ij})$$
$$\mu_{ij} = e_{ij}\theta_{ij}$$
$$\log(\theta_{ij}) = \alpha_0 + S_i + T_j.$$

This type of model often includes a convolution in the spatial term i.e., $S_i = v_i + u_i$, and a simple temporal dependence either as a time trend or as a random walk effect. For a trend we could have a linear function of time such as $T_j : \gamma_j = \beta t_j$ or $\gamma_j \sim N(\beta t_j, \tau)$ where t_j is the time associated with the j th time period (year or month or day etc.). Alternately we could have a more non-parametric random walk dependence such as $\gamma_j \sim N(\gamma_{j-1}, \tau)$. Here three basic models will be examined:

- model 1: a convolution $v_i + u_i$ with $\gamma_j \sim N(\beta t_j, \tau)$

- model 2: a convolution $v_i + u_i$ with $\gamma_j \sim N(\gamma_{j-1}, \tau)$

- model 3: a uncorrelated spatial term v_i with a random walk $\gamma_j \sim N(\gamma_{j-1}, \tau)$

The matrices count[,] and expe[,] are the observed and expected counts for the disease in question, with rows as regions (m) and columns as time periods (T).

Model 1

```
STCode1<-nimbleCode(
{
for (i in 1:m)
{
    for (k in 1:T)
    {count[i,k]~dpois(mu[i,k])
 log(mu[i,k])<-log(expe[i,k])+log(theta[i,k])
log(theta[i,k])<-a0+g[k]+u[i]+v[i]
    }
v[i]~dnorm(0,tauv)
}
for (k in 1:T){
g[k]~dnorm(b0*t[k],taug)
}
u[1:m]~dcar_normal(adj[1:L],weights[1:L],num[1:m],tauu,zero_mean=1)
for(k in 1:L) {weights[k] <- 1}
```

```
a0~dnorm(0,tau0)
tauu~dgamma(2,0.5)
taug~dgamma(2,0.5)
tauv~dgamma(2,0.5)
tau0~dgamma(2,0.5)
b0~dnorm(0,taub)
taub~dgamma(2,0.5)
})
```

Model 2

```
STCode<-nimbleCode(
{
for (i in 1:m)
{
    for (k in 1:T)
    {count[i,k]~dpois(mu[i,k])
 log(mu[i,k])<-log(expe[i,k])+log(theta[i,k])
log(theta[i,k])<-a0+g[k]+u[i]+v[i]
    }
v[i]~dnorm(0,tauv)
}
g[1]~dnorm(0,taug)
for (k in 2:T){
 g[k]~dnorm(g[k-1],taug)
}
u[1:m]~dcar_normal(adj[1:L],weights[1:L],num[1:m],tauu,zero_mean=1)
for(k in 1:sumNumNeigh) {weights[k] <- 1}
a0~dnorm(0,tau0)
tauu~dgamma(2,0.5)
taug~dgamma(2,0.5)
tauv~dgamma(2,0.5)
tau0~dgamma(2,0.5)
})
```

Model 3

```
STCode<-nimbleCode(
{
for (i in 1:m)
{
    for (k in 1:T)
    {count[i,k]~dpois(mu[i,k])
```

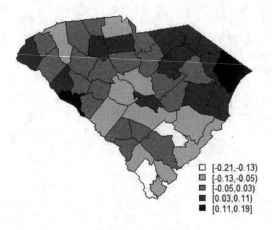

FIGURE 12.1
Posterior mean correlated random effect (CH) for the temporal random walk
model (model 2).

```
log(mu[i,k])<-log(expe[i,k])+log(theta[i,k])
log(theta[i,k])<-a0+g[k]+v[i]
#
    }
v[i]~dnorm(0,tauv)
}
g[1]~dnorm(0,taug)
for (k in 2:T){
 g[k]~dnorm(g[k-1],taug)
}
a0~dnorm(0,tau0)
tauu~dgamma(2,0.5)
taug~dgamma(2,0.5)
tauv~dgamma(2,0.5)
tau0~dgamma(2,0.5)
})
```

Figure 12.1 displays the posterior mean correlated effect (CH) for the model
2, where a convolution is assumed and a random walk in time is included.
Figure 12.2 displays the posterior mean uncorrelated effect (UH) for the same
model.

Figure 12.3 displays the sequence of posterior mean relative risk maps for
the six year period for model 3, with temporal random walk and UH effect
only, in the SCRCST data example.

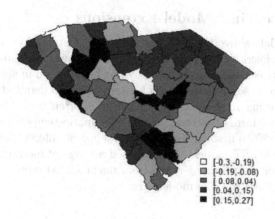

FIGURE 12.2
Posterior mean uncorrelated effect (UH) for the SCRCST data example, for model 2.

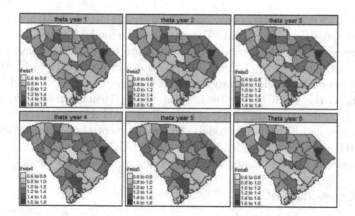

FIGURE 12.3
Posterior mean relative risk estimates for the 6 years in the SC county level respiratory cancers example: for model 3 with random walk in time and uncorrelated spatial heterogeneity.

12.2   Space Time Model Extensions

The basic models above assume that there is no interaction between spatial and temporal effects. This could be a simplistic assumption, and while strong interaction between space and time may not be common in small area health examples, there is some evidence that interactions do manifest themselves in localized clustering in space-time. Hence it is often useful to consider the addition of some form of interaction effect in spatio-temporal models.

Knorr-Held (2000) first suggested the modeling of different forms of additive interaction within ST models. He proposed a range of interaction models for ψ_{ij}. These prior distributions range from uncorrelated noise to sophisticated correlated interaction. The models are:

- model 4: a convolution $v_i + u_i$ with $\gamma_j \sim N(\gamma_{j-1}, \tau)$ and $\psi_{ij} \sim N(0, \tau_\psi)$

- model 5: a convolution $v_i + u_i$ with $\gamma_j \sim N(\gamma_{j-1}, \tau)$ and $\psi_{ij}|_i \sim N(\psi_{i,j-1}, \tau_\psi)$

- model 6: a convolution $v_i + u_i$ with $\gamma_j \sim N(\gamma_{j-1}, \tau)$ and $\psi_{ij}|_j \sim ICAR(\tau_u)$

Model 4 assumes an IID Gaussian prior distribution which allows the interaction to absorb extra/residual ST variation, while model 5 allows for temporal dependence via a random walk in the interaction term. Model 6 instead models the spatial structure of the ST interaction via a conditional ICAR prior distribution on the region component. Of the three interaction models mentioned here, the WAICs were 1984.39 (model 4), 1989.69 (model 5), 1991.02 (model 6), and so it seems for these data all the interaction models have lower WAIC than the uncorrelated models. Of the three correlated interaction models the IID interaction (model 4) has lowest WAIC.

The code and initial values for model 4 are given below:

```
STCode4<-nimbleCode(
{
for (i in 1:m)
{
for (k in 1:T)
{count[i,k]~dpois(mu[i,k])
log(mu[i,k])<-log(expe[i,k])+log(theta[i,k])
log(theta[i,k])<-a0+g[k]+u[i]+v[i]+psi[i,k]
```

```
    psi[i,k]~dnorm(0,taupsi)
    }
theta1[i]<-theta[i,1]
theta2[i]<-theta[i,2]
theta3[i]<-theta[i,3]
theta4[i]<-theta[i,4]
theta5[i]<-theta[i,5]
theta6[i]<-theta[i,6]
psi1[i]<-psi[i,1]
psi6[i]<-psi[i,6]
 v[i]~dnorm(0,tauv)
 }
g[1]~dnorm(0,taug)
for (k in 2:T){
g[k]~dnorm(g[k-1],taug)
}
u[1:m]~dcar_normal(adj[1:L],weights[1:L],num[1:m],tauu,zero_mean=1)
for(k in 1:sumNumNeigh) {weights[k] <- 1}
a0~dnorm(0,tau0)
tauu~dgamma(2,0.5)
taug~dgamma(2,0.5)
tauv~dgamma(2,0.5)
tau0~dgamma(2,0.5)
taupsi~dgamma(2,0.5)
})
###################################
##################### inits 4 ########
LSTinits4<-list(a0=0.1,tauv=0.1,tau0=0.1,taug=0.1,taupsi=0.1,
tauu=0.1,g=c(0,0,0,0,0,0),
v=c(0, 0, 0, 0, 0, 0, 0, 0, 0, 0, 0, 0, 0, 0, 0, 0, 0, 0, 0, 0,
0, 0, 0, 0, 0, 0, 0, 0, 0, 0, 0, 0, 0, 0, 0, 0, 0, 0, 0, 0, 0,
0, 0, 0, 0, 0),
u=c(0, 0, 0, 0, 0, 0, 0, 0, 0, 0, 0, 0, 0, 0, 0, 0, 0, 0, 0, 0,
0, 0, 0, 0, 0, 0, 0, 0, 0, 0, 0, 0, 0, 0, 0, 0, 0, 0, 0, 0, 0,
0, 0, 0, 0, 0),psi=structure(.Data=rep(0.0,276),.Dim=c(46,6)))
```

Figure 12.4 displays the posterior mean estimated relative risks for the 6 years of the SCRCST example for model 4 with IID ST interaction.

Figure 12.5 displays the year one and year six posterior mean ST interaction maps for the Knorr-Held ST interaction IID model. These plots suggest a degree of ST interaction variation over time.

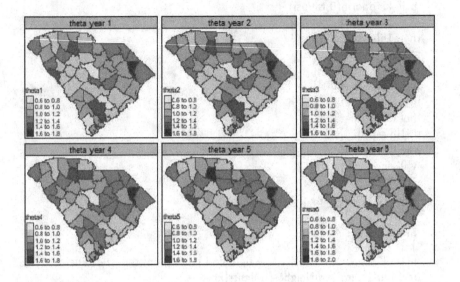

FIGURE 12.4
Posterior mean relative risk estimates for the model 4 with temporal random walk, spatial convolution, and IID Gaussian prior distribution for ψ_{ij}

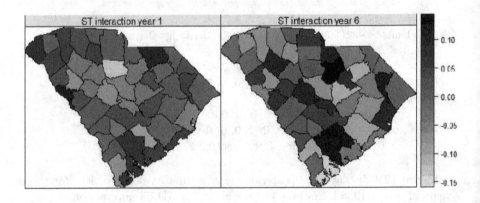

FIGURE 12.5
Knorr-held IID interaction model: SCRCST data example: year 1 and 6 posterior mean ST interaction maps

12.3   Space-time Models in CARBayesST

Many of the ST models fitted using nimble or BUGS can be fitted using CAR-BayesST. Some examples follow. For the SCRCST data example I have fitted models with both Leroux spatial prior specifications and ICAR specification. I have fitted models with separable spatial and temporal components only and with independent interaction, as in the original Knorr-Held interaction specification. Codes for some of these models are given below: S denotes spatial and T denotes temporal random effect prior specifications, and ST denotes an interaction.

1) ### Leroux S and T, no ST ####

```
form<-countL~1+offset(log(expL))
res1<-ST.CARanova(form,family="poisson",W=W.mat, interaction=FALSE,
burnin=10000, n.sample=20000)
```

2) ### Leroux S and T plus ST   (similar to the KH interaction model) ####

```
res2<-ST.CARanova(form,family="poisson",W=W.mat,          interaction=TRUE,
burnin=10000, n.sample=20000)
```

3) ### Leroux S and time trended T no ST ####

```
res3<-ST.CARlinear(form,family="poisson",W=W.mat,burnin=10000,
n.sample=20000,rho.slo=1,rho.int=1,MALA=TRUE)
```

4) ### ICAR S and T + ST ####

```
res4<-ST.CARanova(form,family="poisson",W=W.mat, interaction=TRUE,
rho.S=1,rho.T=1, burnin=10000, n.sample=20000)
```

The resulting goodness of fit for these models is given in Table 12.1. It is noticeable that the Knorr-Held Interaction model with Leroux or ICAR correlation effects and independent interaction prior specification gives the best WAIC values amongst these models. It should be noted that the WAIC results are similar to those found on nimble but it is also important to note that it is not possible to fit a spatial convolution with temporal effect and interaction on CARBayesST as the Leroux model must be used with either a UH-only or ICAR-only alternative to a full Leroux specification.

Model	WAIC	p.eff
1	2011.1	51.8
2	1992.3	69.3
3	97276.0	4759.6
4	1992.4	69.1

Table 12.1
CARBayesST model fitting results for a range of ST models

12.4 An Alternative Modelling Approach: Kalman Filtering

It is common in engineering to model the dynamic systems via two level models describing the system and observation mechanism. Kalman filtering (KF) time series is based on this idea. See Cressie and Wikle (2011) for examples. First, a systems model is specified which governs the underlying behavior of the system, and second an observation model for the system is constructed. In our case we could consider that the relative risk (θ_{ij}) is the parameter of interest and its evolution over time is to be modeled. The observation model is essentially based on counts of disease: y_{ij}. Initially it might be considered that a Poisson observational model could be assumed. However it is possible to mimic the Gaussian KF model if a transformation is used. In particular if the $log(SIR)$ is considered as outcome then it is reasonable to consider an approximate KF model of the general form:

$$log(SIR)_{ij} \sim N(\mu_{ij}, \tau_S^{-1})$$
$$\mu_{ij} = \alpha_0 + \theta_{ij} + S_i + R_j$$
$$log(\theta_{ij}) \sim N(log(\theta_{i,j-1}), \tau_\theta^{-1}\Sigma)$$

where S_i, R_j are spatial and temporal random effects and Σ is a (temporal) covariance matrix. For the simpler $\Sigma = I$ model the code below was fitted to the SCRCST example.

```
Kalman1<-nimbleCode({
for (i in 1:m){
for (k in 1:T){
LSMR[i,k]~dnorm(mu1[i,k],tauS)
mu1[i,k]<-a0+theta[i,k]+Struct[i]+R[k]
}}
for (i in 1:m){
```

```
thetaL[i,1]~dnorm(0,tauTH)
for(k in 2:T){
thetaL[i,k]~dnorm(thetaL[i,k-1],tauTH)
}}
for(i in 1:m){
for (k in 1:T){
log(theta[i,k])<-thetaL[i,k]
}
theta1[i]<-theta[i,1]
theta6[i]<-theta[i,6]
}
for (k in 1:T){R[k]~dnorm(0,tauR)}
###########################################
tauR~dgamma(2,0.1)
Struct[1:m]~dcar_normal(adj[1:L],wei[1:L],num[1:m],tauL,zero_mean=1)
for(k in 1:L){wei[k]<-1}
tauL~dgamma(2,0.1)
tauS~dgamma(2,0.1)
a0~dnorm(0,tau0)
tau0~dgamma(2,0.1)
tauTH<-dgamma(2,0.1)
})
```

Figure 12.6 displays the posterior mean estimates of the year 1 and year 6 θ_{ij}s from this model for a converged sample of 20K size. Figure 12.7 displays the posterior mean estimated temporal effect profile with 95% credible interval, from the converged sample from the KF model fit.

12.5 Some Extensions

Note that these ST models above can be extended in a variety of ways. First, predictor effects can be included to help explain the variation or confounding in the outcome. These predictors could be both static and time varying. For instance $\log(\theta_{ij}) = \alpha_0 + S_i + T_j + x_{ij}^t \beta$ where x_{ij}^t is an element of the matrix of time varying values of x. The regression parameter β can be assumed to have a relatively non-informative zero mean Gaussian prior distribution: $\beta \sim N(0, \tau_\beta^{-1})$. Another question arises as to whether β should be fixed, or time varying, or indeed spatially structured. This could of course arise whether x is spatially or temporally dependent.

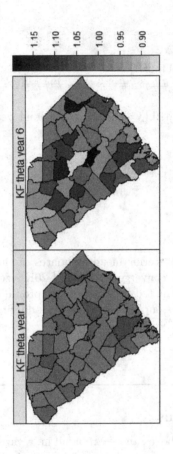

FIGURE 12.6
Posterior mean theta estimates from the Kalman filter space-time model using log(SIR) as outcome.

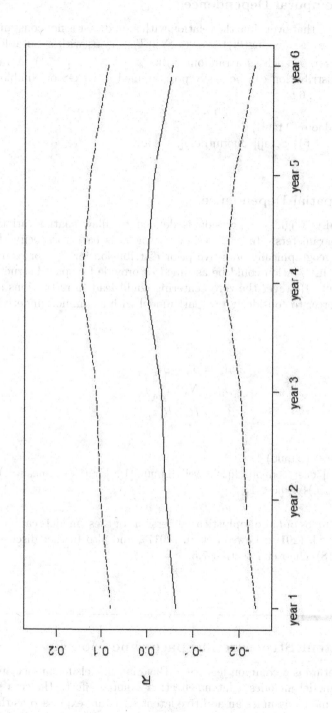

FIGURE 12.7
Posterior mean estimated R profile and upper and lower 95% credible interval from a converged sample from the KF model.

12.5.1   Temporal Dependence

If it is assumed that over time the relation with a predictor is not constant then it is possible to assume $\log(\theta_{ij}) = \alpha_0 + S_i + T_j + x_{ij}^t \beta_j$ with a suitable prior distribution for β_j such as a random walk: $\beta_j \sim N(\beta_{j-1}, \tau_\beta^{-1})$. A random walk prior distribution can be easily programmed in BUGS or Nimble code, for times $j = 1 : 6$:

```
beta[1]~dnorm(0,tauB)
for (j in 2 : 6){ beta[j]~dnorm(beta[j-1],tauB)}
```

12.5.2   Spatial Dependence

Geographically weighted regression is defined to allow spatial variation in regression parameters. In a Bayesian setting it is easy to specify this by assuming a geographically adaptive prior distribution for β_i. For example a ICAR prior distribution could be assumed to provide for spatial structure in the coefficient. However the zero centering could lead to restrictions and so it may be better to consider a two part model with a common underlying β_0 such as:

$$x_{ij}^t \beta_*$$
$$\beta_* = \beta_0 + \beta_{1i}$$
$$\beta_0 \sim N(0, \tau_{\beta_0}^{-1})$$
$$\beta_{1i} \sim ICAR(\tau_{\beta_1}^{-1}).$$

```
bet0~dnorm(0,tau0)
bet1[1:m]~dcar_normal(adj[1:L],wei[1:L],num[1:m],tau1,zero_mean=1)
for(k in 1:L){wei[k]<-1}
```

Some examples of more sophisticated versions of these models can be found in Carroll et al. (2016), Lawson et al. (2017), and also further discussion in Lawson (2018), chapter 7 section 7.5.

12.6   Latent Structure in Space-time Models

Latent structure is a common feature of Bayesian models for in fact any random effect model includes a latent effect: a random effect. Hence a spatial convolution model assumes an additive latent structure expressed as the sum of two random components. However other forms of latent structure can be modeled in particular in the spatio-temporal context. The spatial mixture modeled in chapters 7 and 8 is an example of a fixed mixture model. Essentially this is a simple extension of a convolution model with a mixing probability allowing different areas to have different mixes of components. This type of model can be extended to situations where there are more latent components and also to where there are an unknown number of effects. One approach to this is to allow a large but fixed number of components and to allow the data to select the most relevant. The example in section 11.1.2 is an example of this kind of mixture where areas assign linkage probabilities to different components. Examples of variants of this approach are found in Green and Richardson (2002), Fernandez and Green (2002), and Knorr-Held and Rasser (2000). Another is to allow the number of components to be random and to be estimated. This leads to mixtures with unknown number of components. Reversible jump McMC (RJMCMC) has been proposed to deal with estimation when the dimension of the parameter space is varying Green (1995)). However, it is also possible to simplify this approach by using a large but fixed component space and to use poor man's RJMCMC (Carlin and Chib (1995), Dellaportas et al. (2002)), which leads to some redundant components.

In the spatio-temporal case, latent structure can take on an extended form. A focus of latent modeling could be the disaggregation of temporal effects by spatial area, i.e., spatial clustering of temporal effects/trends. This can be achieved by extending the spatial mixture latent models to include temporal latent components.

Define the relative risk model as

$$\log(\theta_{ij}) = \alpha_0 + \sum_{l=1}^{L} w_{il} \lambda_{lj}$$

where each area has a set of weights w_{il} assigned to L latent temporal components λ_{lj}. This allows the assessment of localized clustering of temporal trends. Lawson et al. (2010) originally proposed the approach and suggested various prior distributional specifications for the resulting models. For the weight matrix a Dirichlet prior distribution is assumed where

$$w_{il} = w_{il}^* / \sum_{k=1}^{L} w_{ik}^*$$

where $w_{ik}^* \sim gamma(1,1)$.

Further modifications of this prior distribution can be made to include spatial correlation for example. In addition a singular multinomial can be assumed which allocates only to the most likely component. When the number of components is unknown a variable selection algorithm can be used to select the components in that

$$w_{il} = \psi_l w_{il}^* / \sum_{k=1}^{L} \psi_l w_{ik}^*$$

with $\psi_l \sim Bern(p_l)$ and p_l could equal 0.5 or be allowed to have a $beta(1,1)$ prior distribution. In this way it is possible to fix L but allow the number of significant components to be $< L$.

The following nimble code was used to model the ST latent structure of the SCRCST data example.

```
STmix1<-nimbleCode({
for (i in 1:m){
for (k in 1:T){
count[i,k]~dpois(mu1[i,k])
log(mu1[i,k])<-log(expe[i,k])+log(theta[i,k])
log(theta[i,k])<-a0+psi[1]*w[i,1]*ch[1,k]+psi[2]*w[i,2]*ch[2,k]+psi[3]*w[i,3]*
ch[3,k]+psi[4]*w[i,4]*ch[4,k]
}
w1[i]<-w[i,1]
w2[i]<-w[i,2]
w3[i]<-w[i,3]
w4[i]<-w[i,4]
#Thmean[i]<-mean(theta[i,1:T])
}
for (k in 1:T)
    {
ch1[k]<-ch[1,k]
ch2[k]<-ch[2,k]
ch3[k]<-ch[3,k]
ch4[k]<-ch[4,k]
}
for (l in 1:4) {
for (k in 1:m){
    w[k,l] <- wstar[k, l] / sum(wstar[k,1:4])
}
delta[1:m,l]~dcar_normal(adj[1:L],wei[1:L],num[1:m],tauL[l])
wstar[1:m,l]<-exp(delta[1:m,l])
tauL[l]~dgamma(1,1)
}
for(k in 1:L){wei[k]<-1}
##############################################
##############################################
#initialise latents: ch[j,1]
for (j in 1 : 4)
{ch[j,1]~dnorm(0,tau.ch[j])
```

```
ch[j,2]~dnorm(0,tau.ch[j])
psi[j]~dbern(0.5)}
# set up time trend in latents ch[j,k]
for (j in 1:4){
for(k in 3:T) {
a[j,k]<-0.6*ch[j,k-1]+0.4*ch[j,k-2]
ch[j,k] ~dnorm(a[j,k],tau.ch[j])
}
}
# Hyperprior distributions on inverse variance parameter of random effects
#priors for tau.ch[j]
for(j in 1 : 4){
tau.ch[j]<-1
}
a0~dnorm(0,tau0)
tau0~dgamma(2,0.5)
})
```

The code assumes a spatially structured prior distribution (ICAR) for the unnormalized weights w^* for each component, and also includes inclusion parameters ψ. The posterior mean estimates for inclusion were: ψ : (0.38810, 1.00000, 0.50225, 0.20215) . This suggests that component 2 has high probability and component 3 less so. Component 1 and 4 are redundant. This is confirmed in Figure 12.8, which show the component 2 widely separated from the others and the other varying around zero. It should be clear that there can be identifiability issues when specifying continuous components such as these and label switching can occur. Posterior assignment checking is recommended (Jasra et al., 2005).

Figure 12.9 displays the posterior mean probability weight map for the second component. It is clear that many areas have high weight on this component but that the north eastern areas load heavily on the component.

Choi and Lawson (2011) provided a simulated evaluation of a range of these models, as well as a Dirichlet Process variant formulation, and Hossain et al. (2014) further evaluated the choices of assignment of weights. Another form of space-time latent clustering is based on the Dirichlet Process and stick breaking prior specification for component weights, and this was further evaluated by Hossain et al. (2012). Most recently the original proposal was used in two stage estimation to reduce bias when spatial random effects and predictors are together in the same model at the same resolution level (Lawson et al., 2012).

To deal with label switching, a discrete version of the latent structure model was proposed by Lee and Lawson (2016) whereby the components are assumed to come from a small set of fixed component types. This avoids the identification issues that can arise when the components are assumed to be stochastic.

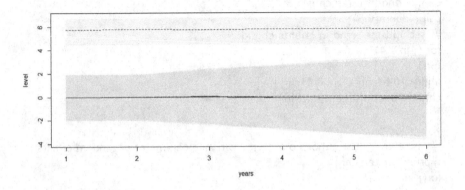

FIGURE 12.8
Posterior mean estimated temporal components, with 95% credible intervals, from a spatially structured weight model and AR2 prior distributions.

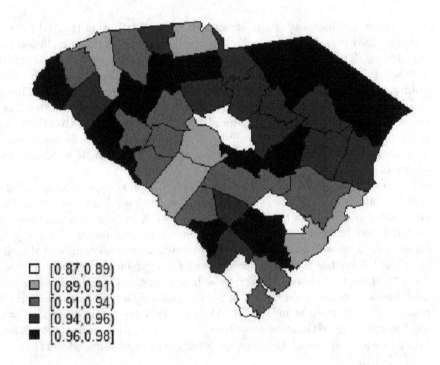

FIGURE 12.9
Posterior mean estimated weight map for second component, from the model with spatially structured weight prior distributions.

For a recent development see Napier et al. (2019). CARBayesST includes a function for this type of model: ST.CARclustrends(). An alternative proposal for a simpler Bayesian model choice approach has subsequently been made by Li et al. (2012): BayesSTDetect.

12.7   Clustering in Space-time Models

Assessment of clustering is an important in small area health studies. Both Public health authorities and epidemiological researchers benefit from being able to assess the degree to which disease cluster. This could suggest localized adverse areas of adverse disease risk. In space -time, clustering can occur in different ways. Spatial clusters can persist, as can temporal clusters across space. Whereas space-Time (ST) clusters can be found to be excesses of risk localised in both space and time.

12.7.1   Exceedence

It is always possible to assess hot spot clusters in small are data using exceedence probabilities. These can be defined for a range of different type of parameters: relative risks, binomial probabilities, spatially structured effects, point process intensities etc. For Poisson data models the relative risk is usually the focus, and so the computation of $\Pr(\theta_{ij} > c)$ is the appropriate estimate. The exceedence probability detects how extreme the value of the parameter is in a given area and time period. This probability describes the upper tail behavior of the distribution of the parameter.

$$\widehat{\Pr}(\theta_{ij} > c) = \frac{1}{G} \sum_{g=1}^{G} I(\theta_{ij}^g > c).$$

As in the spatial case , it is straightforward to program this exceedence within BUGS or nimble code:

```
PRexc[i,j]<-step(theta[i,j]-c)
```

A range of studies have extended these ideas into neighborhood metrics (see, e.g., Hossain and Lawson, 2010).

12.7.2   Modeled Clustering

As in the spatial case, clustering can be modeled. In the spatial case, it is possible to employ partition models whereby areas are classified into areas of similar risk (partitions: see chapter 11). This can be extended into space-time. The modelled risk is assumed to be

$$\log(\theta_{ij}) = \lambda_{z_{ij}} + \phi_{ij}$$

where ϕ_{ij} has a smooth dependent spatio-temporal prior distribution to capture global smoothness, whereas $\lambda_{z_{ij}}$ are a set of G intercept partitions which are allowed to vary in space-time. The intercepts are ordered via the prior distribution, $\lambda_k \sim U(\lambda_{k-1}, \lambda_{k+1})$ $k = 1, ..., G$. The classes are penalized via a temporal and spatial prior distribution $f(z_{ij}|z_{i,j-1})$. The code below can be used to apply ST.CARlocalised() in CARBayesST.

```
library(CARBayesST)
.
.
.
m=46
T=6
######## long form ##########################
countL<-rep(0,276)
expL<-rep(0,276)
for (j in 1:T){
for (i in 1:m){
k<-i+m*(j-1)
countL[k]<-count[i,j]
expL[k]<-expe[i,j]
}}
SCpoly<-readSplus("SC_SPLus_export_map.txt")
plot(SCpoly)
W.nb <- poly2nb(SCpoly)
W.mat <- nb2mat(W.nb, style="B")
##################### model fitting ######
form<-countL~1+offset(log(expL))
res1<-ST.CARlocalised(form,family="poisson",G=6,W=W.mat,burnin=
15000,n.sample=20000,MALA=TRUE)
```

The resulting posterior (median) estimates of lambda are given in Table 12.2.

The WAIC was 2027.50 with pW = 94.1. Figure 12.10 and 12.11 display the label assignments for the different counties for year one and year six from the fitted model for the SCRCST data example. The group assignments overall time were group2 (31) group3 (228) group4 (17). It is clear that the third group is predominant.

Parameter	Median	2.5%L	97.5%L
lambda1	-0.2768	-0.3213	-0.2322
lambda2	0.0904	0.0645	0.1158
lambda3	0.4563	0.3197	0.6158
lambda4	1.0047	0.6177	1.9639

Table 12.2
Posterior median estimates of lambda for the ST.CARlocalised model fitted
to the SCRCST example.

FIGURE 12.10
Posterior mean label assignments for the counties for year one of the fitted
model for the SCRCST example.

FIGURE 12.11
Posterior mean labels for year 6 of the SCRCST data example from the fitted
ST.CARlocalised model.

13

Spatio-Temporal Modeling with INLA

While much space has been devoted to ST modeling with McMC, it is certainly possible to avoid the use of scripted program code, and to employ alternative approximations at least for the simpler ST models. Here I demonstrate a range of models that can be fitted straightforwardly using INLA. The basic assumption of separable spatial and temporal effects is made and a progression of models with increasing complexity is considered. The models for $\log(\theta_{ij})$ are as follows:

$\alpha_0 + v_i$ UH only

$\alpha_0 + v_i + u_i$ convolution

$\alpha_0 + v_i + u_i + \alpha_1 t$ convolution + time trend

$\alpha_0 + v_i + u_i + \gamma_j$ convolution +RW1 time

$\alpha_0 + v_i + u_i + \gamma_j + \psi_{ij}$ Knorr-Held IID interaction model

The code for the model setups is given below. First the data is formatted in long form.

```
countL<-rep(0,276)
expL<-rep(0,276)
for (i in 1:m){
for (j in 1:T){
k<-j+T*(i-1)
countL[k]<-count[i,j]
expL[k]<-expe[i,j]}}
```

Next the graph file is set up from a SPlus export file import.

```
SCpoly<-readSplus("SC_SPlus_export_map.txt")
######## converting adjacency info to graph file ##########
inla.geobugs2inla(adj, num, graph.file="SCgraph.txt")
```

Then the random indices are set up and dataframe created. A separate index is created for the UH and CH effects.

```
t=c(1,2,3,4,5,6)
year<-rep(1:6,len=276)
region<-rep(1:46,each=6)
region2<-region
ind2<-rep(1:276)
STdata<-data.frame(countL,expL,year,t,region,region2,ind2)
```

The following code fits the series of models listed above

13.0.3   Uncorrelated spatial heterogeneity

```
formula1<-countL~1+f(region,model="iid",param=c(2,0.5))
result1<-inla(formula1,family="poisson",data=STdata,
E=expL,control.compute=list(dic=TRUE,cpo=TRUE,waic=TRUE))
```

13.0.4   Spatial convolution model only

```
formula2<-countL~1+f(region,model="iid",param=c(2,0.5))+f(region2,
model="besag",graph="SCgraph.txt",param=c(2,0.5))
result2<-inla(formula2,family="poisson",data=data,
E=expL,control.compute=list(dic=TRUE,cpo=TRUE,waic=TRUE))
```

13.0.5   Spatial convolution+time trend

```
formula3<-countL~1+f(region,model="iid",param=c(2,0.5))+f(region2,
model="besag",graph="SCgraph.txt",param=c(2,0.5))+year
result3<-inla(formula3,family="poisson",data=data,
E=expL,control.compute=list(dic=TRUE,cpo=TRUE,waic=TRUE))
```

13.0.6   Spatial convolution +temporal random walk

```
formula4<-countL~1+f(region,model="iid",param=c(2,0.5))+f(region2,
model="besag",graph="SCgraph.txt",param=c(2,0.5))+
f(year,model="rw1",param=c(1,0.01))
result4<-inla(formula4,family="poisson",data=data,
E=expL,control.compute=list(dic=TRUE,cpo=TRUE,waic=TRUE))
```

Model	WAIC	p.eff
UH only	2008.2	45.60
convolution	2009.1	47.76
convolution +time trend	2011.4	48.90
convolution+RW1 time	2005.0	50.77
convolution+RW1 time+ IID interaction	1983.8	86.40

Table 13.1
Goodness of fit comparisons for a range of spatio-temporal models fitted to
the SCRCST data example using INLA

13.0.7 Spatial convolution +temporal random walk + IID ST interaction (Knorr-Held model I)

```
formula5<-countL~1+f(region,model="iid",param=c(2,0.5))+f(region2,
model="besag",graph="SCgraph.txt",param=c(2,0.5))+
  f(year,model="rw1",param=c(2,0.5))+f(ind2,model="iid",param=c(2,0.5))
result5<-inla(formula5,family="poisson",data=data,
E=expL,control.compute=list(dic=TRUE,cpo=TRUE,waic=TRUE))
```

Table 13.1 displays the goodness-of-fit results of fitting this sequence of models. The Waic is displayed with effective number of parameters. It is clear that for these data the spatial only and time trend models do not differ markedly in their explanation, but the random walk in time does reduce the Waic by over 3 units from the next model. Overall, however the Knorr-Held IID interaction model has a substantially lower WAIC (1983.8) than the other models. It seems that the addition of an interaction to these models enhances the fit considerably. Figures 13.1, 13.2, 13.3, and 13.4 display some posterior mean results from fitting the Knorr-Held IID interaction model to the SCRCST data example. The CH and UH effects are clearly different with the CH effect showing clustering while the UH effect displays a more random and unstructured spatial distribution. The temporal effect displays an increase over the first few years followed by a decrease with the variance of the effect largely constant. Finally, the space-time interaction posterior mean maps can be used to detect transient clustering in space-time, via the examination of areas of excess risk. For example, a marked high in Clarendon county in year 3 is suggestive of an ST cluster. This could be further examined via the use of exceedence probabilities with the computation and examination of $\Pr(\psi_{ij} > 0)$ for each year in turn.

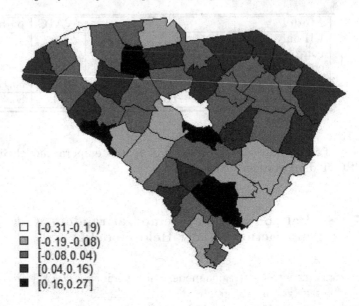

FIGURE 13.1
Uncorrelated effect posterior mean map for the Knorr-Held IID interaction model for the SCRCST data example.

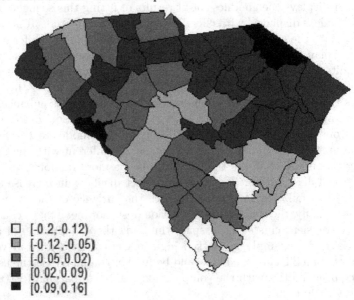

FIGURE 13.2
Correlated (ICAR) effect posterior mean map for the Knorr-Held IID interaction model for the SCRCST data example.

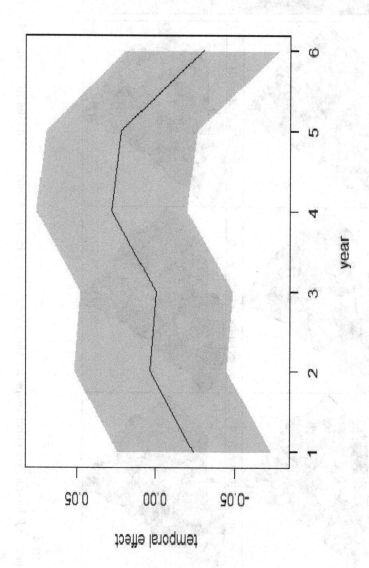

FIGURE 13.3

Posterior mean temporal effect profile with 95% credible interval for the Knorr-Held model fitted to the SRCST data example.

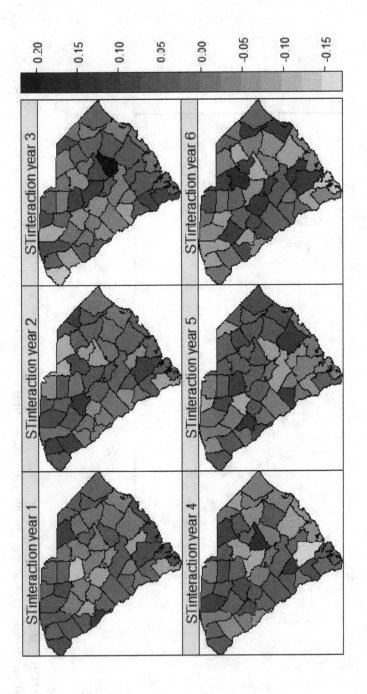

FIGURE 13.4
Posterior mean ST interaction effect maps for the 6 year period for the Knorr-Held model fitted on INLA to the SCRCST data example.

Model	INLA	CARBayesST (5K sample)	CARBayesST (10k sample)	Nimble 5K sample	Nimble 10K sample
ICAR only	6.12	NA	NA	4.58	9.62
KH ICAR STint	2.64	5.34	9.84	9.14	17.36
KH ICAR -STint	2.45	3.48	6.98	6.4	14.45
KH Leroux STint	NA	6.19	11.94	NA	NA

Table 13.2
Computation time (secs) for different packages in the model fitting with the SCRCST data example

13.1 Timing Comparison of Space-time Models in INLA, CARBayes, and Nimble

It is instructive to consider differences in computation time between the different packages, as this could influence users choice of computational platform. Table 13.2 displays the results of timing comparisons. The SCRCST data example was used throughout. This has 276 units with 46 regions and 6 time periods. For two models with ICAR convolution and random walk temporal effect, one with IID interaction (KH model), and one without, system.time() was employed to assess overall time for execution. The INLA models assumed weakly informative precision prior distributions (gamma(2,0.5)) for the spatial effects and a weakly informative prior distribution for the precision of the temporal effect (gamma(1,0.01)). The CARBayesST models assume default precision prior distributions (gamma(0.001,0.001)) and were run for 10000 iterations on a single chain with a 5000 iteration burnin. The Nimble models had precision prior distributions which were also weakly informative (gamma(2,0.5)) with the same run parameters. For these data the McMC apparently converges by 5000 iterations so it should be reasonable to sample beyond a burnin of 5000. Table 13.2 displays the results (in seconds) for the computation on INLA compared to CARBayesST and Nimble for different burnin lengths (5K and 10K). For these data the INLA model fits are usually faster, especially when fitting interaction models. CARBayes appears to be marginally faster than Nimble overall.

Note that this comparison is limited to a relatively small ST data example and so scaling up the times could differ between packages.

Part II

Some Advanced and Special topics

14

Multivariate Models

Often there is a need to consider more than one disease and its spatial distribution. There are various reasons for considering multiple disease distributions. First, the common etiology links could be thought to exist between the diseases and in the spatial patterning of the diseases. Second, disease progression could be represented by the separate diseases. Third, it could be that the links between diseases are to be explored and as yet unknown. An example of the first situation would be examination of diseases commonly affected by air pollution, such as respiratory cancer, bronchitis, COPD, or asthma. While these diseases have various different etiological predispositions, they have some commonality in that they are affected or aggravated by a pollution insult. If that insult is unobserved then the four outcomes could be confounded by this effect. This could imply that a hidden underlying (latent) component should be envisaged.

Examples of the second situation would be stages of disease severity, where each disease is an earlier realization of another. An example of this could be influenza-like illness (ILI) and influenza, where ILI is a precursor of influenza. Another example would be early stage cancer and late stage cancer. An example of the third situation could be the spatial patterning of T1 and T2 diabetes. These are known to have different etiology and affect different age groups but it could be important to assess how related they are via population level analysis.

In the following examples we will focus on the association between asthma incidence and that of chronic obstructive pulmonary disease (COPD) and angina. Both asthma and COPD are respiratory diseases, while angina is a coronary artery condition. Hence, a connection between COPD and angina may be present as they are chronic diseases. The data example is county level incident case counts for asthma, COPD and angina in the state of Georgia USA for 2005. This example appeared in Lawson (2018), Chapter 10 and was introduced in section 1.1.

Basic model notation can be as follows: y_{ik} denotes the count of disease in the i th area/region for the k th disease, for $i = 1, ..., m$ and $k = 1, ..., K$. For the Georgia example $K = 3$: asthma (1), COPD (2) , angina (3). Similarly the expected rates and relative risks are defined as e_{ik}, and θ_{ik}. Data model choice could depend on the nature of the enquiry. Often a Poisson data model is relevant, and conditioning on precedents in the hierarchy, the diseases at the data level could be regradesd as independent Poisson random variables.

Hence assume that

$$y_{ik} \sim Pois(\mu_{ik})$$
$$\mu_{ik} = e_{ik}\theta_{ik}.$$

The ingredients of the models will usually consist of disease specific components and also common components. The different modeling approaches are based on these different dichotomies. First consider two diseases.

14.1   Two Diseases

When there are only two diseases, it is possible to conditionally model a single disease on the total count. Define

$$n_i = y_{i1} + y_{i2}$$
$$y_{i1} \sim bin(p_{i1}, n_i).$$

By conditioning on n_i the resulting model is binomial with probability p_{i1}. Note that this is a form of competing risk model as the probability relates to the chance of disease 1 occurring compared to disease 2. The probability is modeled as

$$\log it(p_{i1}) = \beta_0 + \log(e_{i1}/e_{i2}) + R_i$$

where the ratio of expected rates is included as an adjustment and additional random effects appear in R_i (Dabney and Wakefield, 2005). An alternative to conditioning is to simply allow shared components between the diseases. This approach can be generalized to multiple diseases also. For two diseases a basic model might be

$$y_{i1} \sim Pois(\mu_{i1})$$
$$y_{i2} \sim Pois(\mu_{i2})$$
$$\log(\theta_{i1}) = \alpha_1 + R_{i1} + W_i$$
$$\log(\theta_{i2}) = \alpha_2 + R_{i2} + W_i.$$

Here, the specific components are $\alpha_1 + R_{i1}$ and $\alpha_2 + R_{i2}$, whereas W_i is a shared common random component. Often it is assumed that W_i is spatially structured so that a clustering in the common component is possible. Also we can have a convolution or Leroux /mixture model for the disease specific part: e.g., $R_{i1} = v_{i1} + u_{i1}$. Additional modulation via fixed effect predictors can be added to either log relative risk of course. For example for disease 1 we could have $\log(\theta_{i1}) = \alpha_1 + s(x_i^t\beta) + R_{i1} + W_i$ where $s(x_i^t\beta)$ is a smooth function of the linear predictor $x_i^t\beta$.

Within BUGS or Nimble, these types of models are easy to set up as they allow the specification of *joint likelihood* models (joint models for short). A example of joint model code is given below. This is BUGS code but can be easily adapted to Nimble. For brevity only the main for loop is shown and the two disease outcomes are asthma (asthma[]) and COPD (COPD[]) with expected rates Easthma[] and ECOPD[]. Both asthma and COPD are assumed to have an uncorrelated disease specific random effect and a common spatially correlated effect Wcom[]. Wcom[] is assumed to have an ICAR prior distribution.

```
for(i in 1:N){
asthma[i]~dpois(Muas[i])
log(Muas[i])<-log(Easthma[i])+log(thetAS[i])
log(thetAS[i])<-b0+Vas[i]+Wcom[i]
COPD[i]~dpois(MuC[i])
log(MuC[i])<-log(ECOPD[i])+log(thetCOPD[i])
log(thetCOPD[i])<-b1+VaC[i]+Wcom[i]
Vas[i]~dnorm(0,tVas)
VaC[i]~dnorm(0,tVac)
}
for( i in 1:sumNumNeigh){weights[i]<-1}
Wcom[1:N]~car.normal(adj[],weights[],num[],tWcom)
```

Wcom[], Vas[], and VaC[], as well as the relative risk parameters: thet-COPD[] and thetAS[], can all be mapped and examined for etiological evidence.

14.2 Multiple Diseases

Extending the idea of shared components within joint models is straightforward to more than two diseases. However a different possibility is also available. It is possible to consider a truly multivariate model for the effects between the diseases. Such a model assumes a multivariate distribution for the underlying disease variation. It also provides for an inter-disease (cross) correlation, which is a summary measure of similarity between the diseases. This is not available within the shared component formulation. Models considered here are multivariate CAR models (both MICAR and the proper MPCAR), M-models, and the factor models discussed in Chapter 11. Multivariate CAR models have been developed by Gelfand and Vounatsou (2003) and further by Jin et al. (2005) and Jin et al. (2007). A basic formulation assumes that the spatially structured random effect has a multivariate CAR model defined by:

$$\log(\theta_{ki}) = \phi_{ki} + \text{other terms}$$
$$\phi_i \sim MCAR(\alpha, \Lambda)$$

where $MCAR(\alpha, \Lambda)$ is defined to be a multivariate normal distribution as $N_{mK}(\mathbf{0}, \{\Lambda \otimes (D - \alpha W)\}^{-1})$ where Λ is a $K \times K$ nonspatial precision matrix between diseases and $\otimes$ is a Kronecker product. $(D - \alpha W)$ is the kernel of spatial univariate CAR model with binary weight matrix W and D is diagonal matrix with entries the number of neighbors of each area/region. α is the spatial association parameter. It is assumed to be constant across diseases. When $\alpha = 1$ this becomes $MCAR(1, \Lambda)$, a multivariate ICAR model (MICAR). The original $MCAR(\alpha, \Lambda)$ model is known as multivariate PCAR (MPCAR) with common α parameter. A common parameter may seem inflexible, and an extension to allow separate parameters has also been proposed by Jin et al. (2007). On BUGS the MICAR is preprogrammed as the mv.car distribution. However it is possible to program both the MPCAR and the vector α parameter variant in either BUGS or, more efficiently, in Nimble. As far as this author is aware these are not available *directly* in INLA or CARBayes.

The code for the MPCAR nimble model is given below. Using the adj and num vectors and converting to the C and M matrices, bounds for the PCAR parameters are derived using the carBounds() function:

```
CM2<-as.carCM(adj[1:L],weights[1:L],num[1:Nareas])  .
C<-CM2$C
M<-CM2$M
bou<-carBounds(C,adj,num,M)
if(bou[1]<0)bou[1]=0
R = structure(.Data = c(1, 0, 0, 0, 1, 0, 0, 0, 1), .Dim = c(3,3))
Q = structure(.Data = c(1, 0, 0, 0, 1, 0, 0, 0, 1), .Dim = c(3,3))
K<-Ndiseases
###############################################
kl<-K*Nareas
LSTConsts<-list(C,M,bou,Nareas,Ndiseases,mumod,E,num,adj,sumNumNeigh,
L,R,K,kl)
names(LSTConsts)=c('C','M','bou','Nareas','Ndiseases','mumod','E','num','
adj','sumNumNeigh','L','R','K','kl')
#attach(LSTConsts)
mumod<-rep(0,length=Nareas)
###############################################
##### models for Georgia 3 diseases ################
### Multi independent spatial effects with PCAR models#########
##### This is the full MCAR(alpha,Cov) model of Jin et al (2007)
##### denote this as the MPCAR model ###############
```

```
MpCARcode<-nimbleCode(
{
for (i in 1:Nareas)
{
for (k in 1:Ndiseases)
    {O[i,k]~dpois(mu[i,k])
Opred[i,k]~dpois(mu[i,k])
 log(mu[i,k])<-log(E[i,k])+log(theta[i,k])
log(theta[i,k])<-a0[k]+phi[i,k]+v[i,k]
v[i,k]~dnorm(0,tauv[k])
Dp[i,k]<-(O[i,k]-Opred[i,k])**2
}
}
for (k in 1:Ndiseases){
a0[k]~dnorm(0,tau0[k])
tauv[k]~dgamma(2,0.5)
tau0[k]~dgamma(2,0.5)
tauu[k]<-1
gamma[k]~dunif(bou[1],bou[2])
u[1:Nareas,k]~dcar_proper(mumod[1:Nareas],C[1:L],adj[1:L],num[1:Nareas],
M[1:Nareas],tauu[k],gamma[k])
}
Prec[1:K,1:K]~dwish(R[1:K,1:K],K)
Cov[1:K,1:K]<-inverse(Prec[1:K,1:K])
Achol[1:K,1:K]<-chol(Cov[1:K,1:K])
for(i in 1:Nareas){
phi[i,1:K]<-Achol[1:K,1:K]%*%u[i,1:K]}
sig[1]<-sqrt(Cov[1,1])
sig[2]<-sqrt(Cov[2,2])
sig[3]<-sqrt(Cov[3,3])
cor12<-Cov[1,2]/(sig[1]*sig[2]) # between asthma and COPD
cor13<-Cov[1,3]/(sig[1]*sig[3]) # between asthma and angina
cor23<-Cov[2,3]/(sig[2]*sig[3]) # between COPD and angina
mspe<-mean(Dp[1:Nareas,1:Ndiseases])
})
```

14.2.1   M-models

In a sequence of papers a different formulation of multivariate disease variation
has been proposed by Botella-Rocamora et al. (2015), and Martinez-Beneito
et al. (2017). In these models a combination of spatially structured factor
effects are assumed to underlie the risk surfaces of groups of disease. These
effects can have proper CAR prior distributions or ICAR prior distributions.

The vector of log relative risk is posited to be described by

$$\phi_k = \kappa_1 m_{1k} + \ldots\ldots \kappa_m m_{mk}$$

where κ_i is a spatial MRF (either PCAR or ICAR) and m_{ik} is the i, k th element of a matrix M, which is a weight matrix. The elements of M are often assumed to have zero mean Gaussian distributional form with large variances which are relatively non-informative, or large range uniform. The number of components does not have to be the same as the number of areas and indeed could be less or more depending on the application. Identification of the components in this mixture is problematic however, as noted by the authors.

14.2.2   Dimension Reduction

In Chapter 11, an example of dimension reduction was given (see section 11.2). It is possible to hypothesize that an underlying single spatially structured factor effect is present in multivariate disease data. This assumes a disease specific factor loading to a common factor i.e.,

$$\mu_{ik} = e_{ik}\rho_{ik} \qquad (14.1)$$
$$\log(\rho_{ik}) = \lambda_k f_i$$

That model fitted to the Georgia three disease example yielded posterior mean estimates of λ : (0.15491786, 0.31680317, 0.43252911) for asthma, COPD, and angina, respectively, and the posterior mean estimate of f_i is displayed in Figure 11.7.

14.2.3   Comparison of Models

It is instructive to consider how well different models perform in terms of goodness of fit for particular data examples. The models fitted were: (1) the multivariate factor model of Wang and Wall (2003) (see section 11), (2) a multivariate ICAR model with a Wishart prior distribution for the precision matrix Λ, (3) multivariate PCAR model ($MCAR(\alpha, \Lambda)$) with a Wishart prior distribution for the precision matrix Λ, and uniform prior distribution for the α vector, (4) separate convolution models for each disease, (5) shared spatially structured (ICAR) component with separate disease specific convolution models, (6) as for (5) but with only uncorrelated disease - specific effects, (7) an M-model using ICAR base models. In the case of the Georgia three disease data, Table 14.1 displays the results of nimble WAIC computation (see section 5.2) and predictive fit (MSPE). It is clear that while a shared component model, with or without disease specific random spatial effects, has low WAIC, the proper MCAR (MPCAR) model provides the overall lowest WAIC for these data. Separate convolution models are not competitive and the M-model fitted with 6 ICAR components is slightly worse, while the MVFA model is considerably poorer in terms of WAIC (5548.07). The MSPE

Model	WAIC	MSPE (sd)
MVFA	5548.07	733.44 (55.33)
MICAR	3141.29	118.05 (16.38)
MPCAR	3133.75	116.66 (15.71)
MV separate convolution models	3152.43	118.44 (16.74)
MV shared component with convolutions	3138.74	119.29 (17.13)
MV shared component UH specific only	3140.41	117.56 (15.90)
M-models	3142.27	117.22 (16.10)

Table 14.1
Comparison of seven different multivariate models for the Georgia 3 disease data example

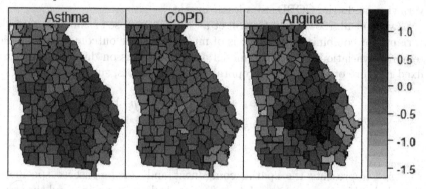

FIGURE 14.1
Posterior mean maps of the PCAR spatial effects for the Georgia 3 disease example.

results largely mimic the WAIC, though less differentiated, with the MPCAR being the lowest in terms of MSPE. However most models have MSPE below 120, while the MVFA model has a much higher MSPE.

Figure 14.1 displays the posterior mean estimated proper CAR effects for the MPCAR model. The posterior mean estimates of the spatial association parameters (γ) were: asthma: 0.904 (0.117), COPD: 0.582 (0.267), angina: 0.951 (0.037).

14.3 Multivariate ST (MVST) Models

14.3.1 General MVST Models

It is possible to extend the models considered in previous sections into the spatio-temporal domain. Once time is admitted this extends the focus considerably. The form of extension could determine the form of model considered. However it should be noted that other forms of time dependence could be encountered, such as where time is observed randomly and different spatial units are observed at different times. Alternatively there could be data on duration of temporal effects, where a spatial unit might have a temporal extent that differs from other units. Here I will simply consider fixed time periods and fixed spatial units.

Define the outcomes as y_{ijk} , $i = 1, ..., m$, areas, $j = 1, ..., J$,time units and $k = 1, ..., K$ diseases.

Figure 14.2 displays the SIR maps for the three diseases in Georgia counties over the 8 years (2011-2018). The common legend reduces the asthma SIRs, but it is clear that the angina SIRs display marked spatial structure and time variation as do the COPD, to a lesser extent.

Lawson et al. (2017) and Carroll et al. (2017) proposed a number of model variants for combining the analysis of multiple disease outcomes with spatio-temporal variation. The basic model assumed a Poisson data model with a fixed mixture of spatial and temporal risk components, such as

$$\log(\theta_{ijk}) = \alpha_k + p_{ik}M_{ik}^S + (1 - p_{ik})M_{ijk}^{ST}$$
$$M_{ik}^S = v_{ik} + u_i$$
$$M_{ijk}^{ST} = \gamma_j + \psi_{ijk}.$$

The M_{ik}^S represents the spatial component and can be disease specific or shared. In this case an uncorrelated effect v_{ik} is disease specific and the structured effect u_i is common to all. The second component is spatio-temporal and subsumes the temporal effects, if any. In this case there is a common temporal effect γ_j and a unit specific space-time interaction effect ψ_{ijk}. The linking probability (p_{ik}) is assumed to be only location and disease specific, but can have spatial structure, and in the models considered it has a logistic-link ICAR specification. Variants of these models can be developed and they are discussed in more detail in Lawson et al. (2017). For an example of lung and bronchus, oral-pharyngeal and melanoma cancer space-time data in South Carolina, figure 14.3 demonstrates the difference in spatial effects (v_{ik} and u_i) for OCPCa when univariate, bivariate (OCPCa and LBCa), and full multivariate models are assumed.

In addition to general fixed mixture models it is possible to assume unknown numbers of components in the mixture and this leads on to the latent structure MVST models. Also more specific effects, where known dependencies between diseases are to be included, could be imagined. A classic example would be lagged effect between a pre-cursor primary disease and a secondary disease. An example would be influenza-like illness (ILI) and influenza, or mild cognitive impairment (MCI), and Alzheimer's (AD) disease. In a zoonotic disease study, an animal infection could be observed before a human infection (see e.g., Rotejanaprasert et al. (2017), Lawson and Rotejanaprasert, 2018).

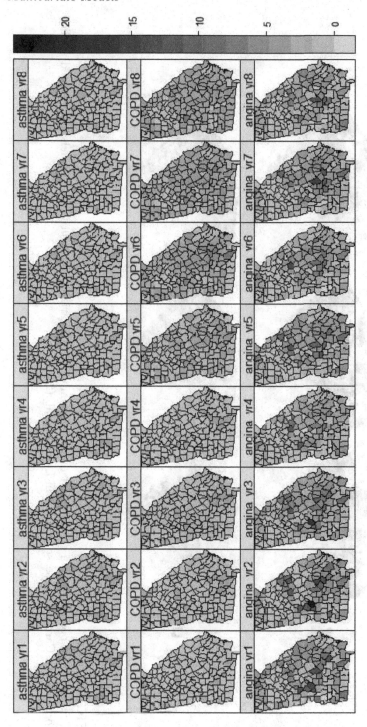

FIGURE 14.2

SIR maps of the three diseases in the Georgia county level MVST data example : asthma, COPD, angina over eight years (2011–2018).

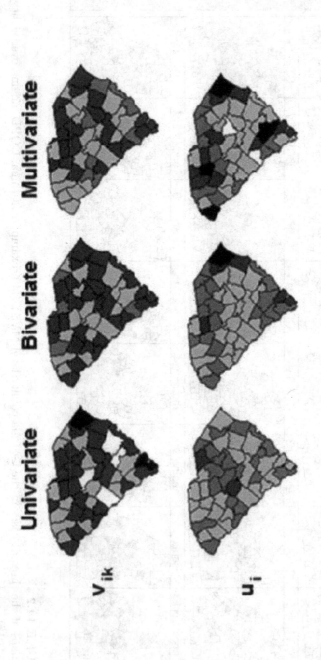

FIGURE 14.3
Comparison of posterior mean spatial effect maps when univariate, bivariate, and multivariate models are assumed applied to the South Carolina 3 disease example in Lawson et al. (2017).

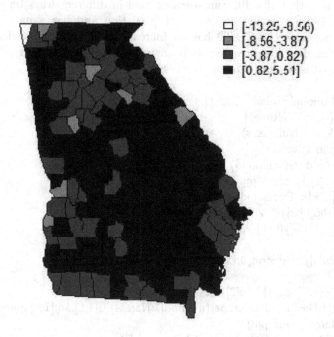

FIGURE 14.4
Posterior mean estimate of the spatial factor for the MVSTFA model applied
to the Georgia three disease ST data example.

14.3.2 Latent Structure MVST Models

14.3.2.1 MVFAST Models

The original model of Wang and Wall (2003) can be extended in time in a
number of ways. The simplest extension with MVST data is to assume that
the loading is time dependent. This leads to a model of the form:

$$\mu_{ijk} = e_{ijk}\rho_{ijk} \tag{14.2}$$
$$\log(\rho_{ijk}) = \lambda_{jk}f_i.$$

with either an unstructured linking: $\lambda_{jk} \sim N(0, \tau_\lambda^{-1})$ or a dependent AR1
or random walk linking: $\lambda_{jk} \sim N(\lambda_{j-1,k}, \tau_\lambda^{-1})$.

The following code is for the unstructured MVSTFA model, but the AR1
version is straightforward adaptation. The posterior mean estimates of the
λ_{jk} were found to be Figure 14.4 displays the resulting spatial factor effect

Figure 14.5 displays the posterior mean temporal profiles for the linking
parameter for each disease to the spatial factor.

It is clear that the different diseases load in different ways on the spatial factor. Asthma profile displays little variation and has a narrow credible interval whereas the COPD has an increasing linkage (and wider interval) whereas angina is decreasing with an interval of greater width. The code model used is displayed below:

```
FAT1model<-nimbleCode({
for ( j in 1: Ntime){
for(k in 1:Ndiseases){
for(i in 1:Nareas){
y[i,j,k]~dpois(mu[i,j,k])
ypred[i,j,k]~dpois(mu[i,j,k])
mu[i,j,k]<-Expe[i,j,k]*rho[i,j,k]
log(rho[i,j,k])<-lambda[j,k]*fact[i]
Dp[i,j,k]<-(y[i,j,k]-ypred[i,j,k])**2
}
lambda[j,k]~dnorm(0,tauL)
}}
phi~dunif(bou[1],bou[2])
fact[1:Nareas]~dcar_proper(mumod[1:Nareas],C[1:L],adj[1:L],num[1:Nareas],
M[1:Nareas],tauF,phi)
  mspe1<-mean(Dp[1:Nareas,1:Ndiseases,1])
  mspe2<-mean(Dp[1:Nareas,1:Ndiseases,2])
  mspe3<-mean(Dp[1:Nareas,1:Ndiseases,3])
  mspeT<-mspe1+mspe2+mspe3
  tauL~dgamma(2,0.5)
  tauF~dgamma(2,0.5)
})
```

Tzala and Best (2008) extend the MVFA models with a variety of temporal extensions. In particular they examine a range of models with the general definition

$$\log(\mu_{ijk}) = \beta_k x_{ij} + \lambda_{sk} f_{si} + \lambda_{jk} f_{tj} + u_{ki} + \gamma_{kj}$$

where f_{si} is a spatial factor and f_{tj} a temporal one. $u_{ki} + \gamma_{kj}$ is a combination of spatial and temporal effects. The above code can be extended to accommodate this extension and other variants.

14.3.2.2   Mixture MVST

Recently, a coupled multivariate ST latent structure model has been used to model progression from mild cognitive impairment (MCI) to Alzheimer's disease (AD). Baer et al. (2020) proposed a model which links the latent ST structure of the two diseases. The latent structure consist of temporal relationship, $\beta_2 = c \rho \beta_1$ or $\beta_1 = c \rho^{-1} \beta_2$. This is a convenient method for examining the correlation between two diseases' profiles. The example includes 3 diseases and so only . . .

$$\ldots$$

$$\ldots$$

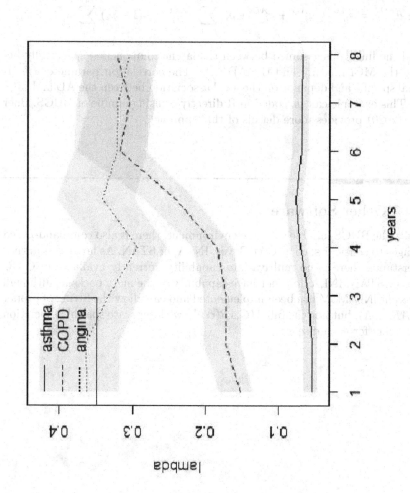

FIGURE 14.5
Posterior mean λ profiles for the three diseases for the MVSTFA model with associated 95% credible intervals.

components which are weighted for each area. These components can take a variety of forms and can also be common between two or more diseases. The example below is for MCI and AD only.

$$\log(\theta_{ij}^{MCI}) = \alpha_0^{MCI} + u_i^{MCI} + v_i^{MCI} + \sum_{l=1}^{L} \Lambda_{lij}^{MCI} + \psi_{ij}^{MCI}$$

$$\log(\theta_{ij}^{AD}) = \alpha_0^{AD} + u_i^{AD} + v_i^{AD} + p_i \sum_{l=1}^{L_{MCI}} \Lambda_{lij}^{MCI} + (1 - p_i) \sum_{l=1}^{L_{AD}} \Lambda_{lij}^{AD} + \psi_{ij}^{AD}$$

and the linkage is assumed between the latent components, specifically between the MCI mixture and the AD risk. The correlation parameter (p_i) is spatial specific and depends on the local association between the AD and MCI risk. This example can be coded in R directly or using nimble or BUGS. Baer et al. (2020) provides more details of this approach.

14.4   Other Software

Besides the BUGS/nimble software environment, there is also command-based packages to consider such as CARBayes, INLA, or STAN. As far as this author understands there is no multivariate capability currently available in CARBayes or STAN. INLA does not have capability in the main package, although the recent INLAMSM has been implemented and can allow the fitting of proper $MCAR(\alpha, \Lambda)$, but not the full $MCAR(\alpha, \Lambda)$ with separate spatial association parameters for each disease.

15

Survival Modeling

It is often the case that the outcome of interest is not in the form of an incident count but is a *time* to an end-point. While observed within a spatial and temporal unit the occurrence of disease naturally falls into a count format, at the individual level there is a diagnosis date and time. By aggregating to a spatial and temporal unit the exact diagnosis date for an individual is lost and only the occurrence within the unit of interest is recorded. If the diagnosis date is available at the patient (individual level) then the focus of analysis is changed. When such time based data is available then time itself is the random quantity. The methods commonly known as *survival analysis* address data in the form of time to end-point. A range of examples of the application of Bayesian spatial modeling to survival data are available (e.g., Osnes and Aalen, 1999, Henderson et al., 2002, Banerjee et al., 2003, Cooner et al., 2006, Carlin and Banerjee, 2003, Diva et al., 2008, Bastos and Gamerman, 2006, Zhou et al., 2008, Onicescu et al., 2017a, and for a review Banerjee, 2016, Fisher and Lawson, 2020). At the population level, a common example of end point data arises from cancer registries, where any occurrence of cancer is logged for patients with date of diagnosis and related demographic and disease-specific outcomes. In this chapter, I will examine a large dataset of prostate cancer (PrCa) from the SEER registry for Louisiana. The data are for the years 2007–2010 and consists of 10246 de-identified subjects with the following outcome and predictor variables: time to vital outcome (end-point), stage at diagnosis (localized/regional(0) or distant (1)), race (Black (1) versus White/other(2)), marital status (1/2/3), and age at diagnosis. Geocoding to county level is the spatial reference.

Censoring occurs within these data as many subjects do not have a vital outcome and so their end point is a censoring time (right censoring). This has an implication for the modeling of the times. A status variable (0/1) is also included in this example which acts as a censoring indicator. Left censoring may also occur but we make the assumption that for all those included between the dates there are exact times or right censored times.

15.1   Some Basic Survival Concepts and Functions

Usually it is assumed that the random time t is governed by a density $f(t)$ and we have a sample of times (end-points) t_i $i = 1,...N$. The times are censored via a censoring indictor $\delta_i = \begin{cases} 0 \text{ censored} \\ 1 \text{ exact} \end{cases}$. Then for each person we observe (t_i, δ_i). The distribution of times is also related to a survivor function $S(t)$ which represents the probability of surviving until t: $S(t) = \int\limits_{t}^{\infty} f(x)dx$. The hazard function is associated with these in that it describes the probability of reaching the end point, conditional on having survived up to t i.e. $h(t) = f(t)/S(t)$. These functions can be specified within a Bayesian model for survival. As a preliminary exploratory tool a non-parametric estimator of the survivor function is often examined. The Kaplan -Meier (KM) estimator is available in the survival package in R. The following R code generates such an estimator, plotted in Figure 15.1.

```
PrCaSEER<-data.frame(time,status,race,stage)
fit <- survfit(Surv(time, status) ~race, data =PrCaSEER)
plot(fit, lty = 1:3)
legend(" bottomleft",legend=c(" Race:  Black"," Race:  White "),lty=1:3)
```

The KM plot suggests that there is a racial effect in survival with blacks apparently having poorer long term survival.

15.1.1   Right Censored Data Likelihood

For a Bayesian geospatial analysis we need to consider a likelihood and consider how to incorporate spatial effects within the formulation. Often a parametric form is assumed for $f(t_i)$ and a variety of candidate distributions are available. One common choice for this is the Weibull distribution with density

$$f(t) = \lambda v t^{v-1} \exp(-\lambda t^v) \qquad t \geq 0.$$

This has great flexibility and can model a variety of forms (peaked or monotonic decline, for example). The survivor function is relatively simple also and is defined as $S(t) = \exp(-\lambda t^v)$. The shape of the distribution is defined by v. The scale parameter λ is often used to parameterize the distribution to allow for covariate effects and also to include random components. Hence the distribution becomes, for the i th individual, $f(t_i) = \lambda_i v t^{v-1} \exp(-\lambda_i t^v)$ and the survivor function $S(t_i) = \exp(-\lambda_i t^v)$. With censored data the contribution from exact and censored times must be considered. Exact times contribute to the likelihood via $f(t_i)$ with (right) censored observations contributing via $S(t_i)$. Hence for the data (t_i, δ_i) we have the likelihood:

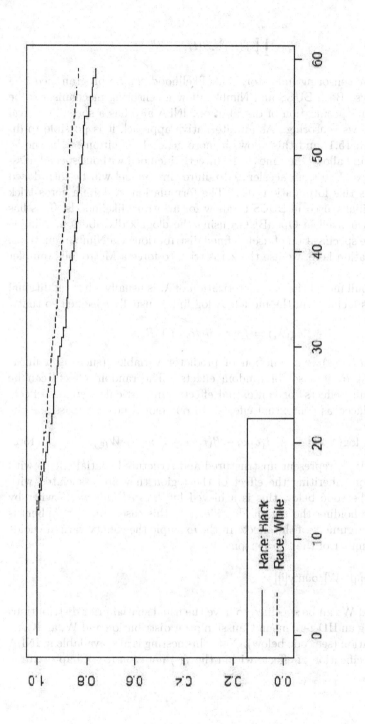

FIGURE 15.1
Kaplan-Meier survivor plot estimate broken by race for the PrCa SEER Louisiana registry example.

$$L = \prod_{i=1}^{N} f(t_i)^{\delta_i} . S(t_i)^{(1-\delta_i)} \qquad (15.1)$$

where δ_i is the censoring indicator. This likelihood can be programmed in a variety of ways. Both BUGS and Nimble allow a censoring mechanism to be included in the specification of distribution. INLA also has a special survival setup that allows censoring. As an alternative approach it is possible to directly program 15.1, and this allows a more general definition of the model components, and allows the times to be directly included without special missing data setups. Later, an accelerated failure time model will be introduced which exploits this formulation also. The formulation exploits a zeros trick that was originally used in BUGS to allow for arbitrary likelihoods. This has since been formalized in OpenBUGS using the dlogLik distribution. Nimble also allows the specification of user defined distributions via Nimble functions. In the formulation here, we use the zeros trick to force a Metropolis sampler algorithm.

In the Weibull model the parameterization of λ_i is usually where additional information is included in the model. A log link is usually specified so that

$$\log(\lambda_i) = f_i(predictors) + R_i$$

where $f_i(predictors)$ is a function of predictor variables (such as a linear predictor) and R_i is a set of random effects. The random effects can be individual frailty effects or contextual effects. In particular, spatial effects can be introduced as contextual effects. In our model below we assume the following:

$$\log(\lambda_i) = \beta_0 + \beta_1 age_i + \beta(race_i) + v_{i \in j} + W_{i \in j}. \qquad (15.2)$$

where $v_{i \in j} + W_{i \in j}$ represent unstructured and structured spatial effects with the i th person inheriting the effect of the region they are associated with ($i \in j$). In the code below this is achieved by "*nested indexing*," whereby a label vector holding the region identifier (in this case, county number) is nested in the argument of the effect. In the example the county variable holds the county number of the i th person:

```
v2[county[i]]+W[county[i]]
```

Both v2 and W can be specified to have the usual spatial prior distributions with v2 having an IID zero mean Gaussian prior distribution and W, an ICAR prior specification (see code below). Note this nesting is also available in INLA simply by specification of effects within the $f(.)$ notation (see Chapter 10).

15.2   Accelerate Failure Time (AFT) Model

A different approach is sometimes taken to modeling survival data. The Weibull distribution is a proportional hazards model and also there are some associated interpretational issues with such a model, especially when mediator effects are to be considered (VanderWeele, 2011, VanderWeele and Tchetgen, 2017). Instead it is possible to consider a model that directly relates the log of time to explanatory variables or effects. The accelerated failure time model makes the following assumptions. The outcome is assumed to be of the form: $log(t)$, and any modulation by predictors or effects act directly on the log of time. The linear predictor for the AFT model is

$$log(t_i) = \mu + x_i^t \beta + R_i + \sigma \varepsilon_i$$

where $x_i^t \beta$ is a linear combination of covariates, R_i are random effects and the final term is a scaled error. The error ε_i arises from a chosen density for the distribution of risk. The density of the time variable is $f(t)$ and the error density is $f_0(.)$. Define $\lambda_i = \mu + x_i^t \beta + R_i$. Hence the density of ε_i has argument $(log(t_i) - \lambda_i)/\sigma$ so that

$$f(t_i) = \frac{1}{\sigma t_i} f_0 \left(\frac{log(t_i) - \lambda_i}{\sigma} \right).$$

The survivor function follows as

$$S(t_i) = S_0 \left(\frac{log(t_i) - \lambda_i}{\sigma} \right)$$

and the likelihood is, as before

$$L = \prod_{i=1}^{N} f(t_i)^{\delta_i} S(t_i)^{(1-\delta_i)}.$$

However, a range of different models can be chosen for $f_0()$ or $S_0()$. In the example below we have used a logistic model for $S_0()$ so that

$$S(t) = \frac{1}{1 + (t_i \exp(-\lambda_i))^{1/\sigma}}.$$

15.3   SEER PrCa Louisiana Example

For this example Nimble was used and an explicit use was made of the censored data likelihood (15.1). In the following code a zeros trick is used and the spatial effects are included via nested indexing.

```
WeibCode<-nimbleCode({
C<-10000
for (i in 1:Nsubj){
log(lambda[i])<-beta0+beta1*age[i]+beta2[race[i]]+v2[county[i]]+W[county[i]]
log(s[i])<-lambda[i]*pow(time[i],nu)
log(f[i])<-log(nu+0.001)+log(lambda[i]+0.001)+(nu-1)*log(time[i]+0.001)-
log(s[i])
L[i]<-exp(status[i]*log(f[i])+(1-status[i])*log(s[i]))
    munew[i]<- -log(L[i])+C
    zeros[i]~dpois(munew[i])      }
beta0~dnorm(0,tau0)
beta1~dnorm(0,tau1)
for (j in 1:2){beta2[j]~dnorm(0,taub[j])
taub[j]~dgamma(1,0.05)}
tau0~dgamma(1,0.05)
tau1~dgamma(1,0.05)
nu~dgamma(1,0.05)
for(i in 1:regions){v2[i]~dnorm(0,tauV1)}
W[1:regions]~dcar_normal(adj[1:sumNN],weights[1:sumNN],num[1:regions],
tauU,zero_mean=1)
for(k in 1:sumNN){weights[k]<-1}
tauV1~dgamma(2,0.5)
tauU~dgamma(2,0.5)
})
```

Note that this model includes age and race as predictors. Figures 15.2 and 15.3 display the posterior mean contextual (UH and ICAR) random effects for the Weibull model fit.

The AFT model code for the logistic distribution is as follows:

```
AFTcode<-nimbleCode({
C<-10000
for(i in 1:Nsubj) {
temp[i] <-(log(time[i]+0.01)-beta0-beta1*age[i]-beta2[race[i]]-W[county[i]])/
sigma
  #survival distribution and density function of logistic distribution
  log(s[i])<-log(1+exp(temp[i]))
  log(f[i])<-2*log(s[i])+temp[i]
  #Loglikelihood Function
    L[i]<-status[i]*log(f[i]/(sigma*(time[i]+0.01)))+(1-status[i])*log(s[i])
    #poisson zero trick
    munew[i]<- -L[i]+C
    zeros[i]~dpois(munew[i])
```

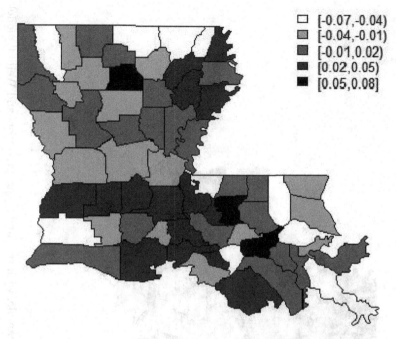

FIGURE 15.2
Posterior mean uncorrelated effect (*v*2) for the Louisiana SEER PrCa data
example, with race and age as predictors and Weibull model.

```
    }
    #spatial from independent normal distribution plus the car model effects
    for(j in 1:sumNN) { weights[j] <- 1}
    W1[1:regions]        ~dcar_normal(adj[1:sumNN],       weights[1:sumNN],
num[1:regions], tauw1,zero_mean=1)
    for (j in 1:regions){ W2[j] ~dnorm(0, tauw2)
    W[j]<-W1[j]+W2[j]}
    #Parameter Prior for the parameters in the AFT model
    beta0 ~dnorm(0.0,0.001)
    beta1~dnorm(0,tau1)
    for(i in 1:2) {beta2[i] ~dnorm(0.0, taub2)}
    taub2~dgamma(2,0.5)
    tau1~dgamma(2,0.5)
    tauw1~dgamma(2,0.5)
    tauw2~dgamma(2,0.5)
     sigma~dgamma(1,0.01)
        })
```

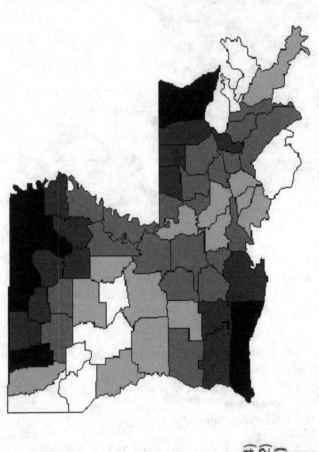

[-0.07, -0.04)
[-0.04, -0.02)
[-0.02, 0.01)
[0.01, 0.03)
[0.03, 0.06]

FIGURE 15.3
Posterior mean ICAR effects for the Louisiana SEER PrCa data example, with race and age as predictors and Weibull model.

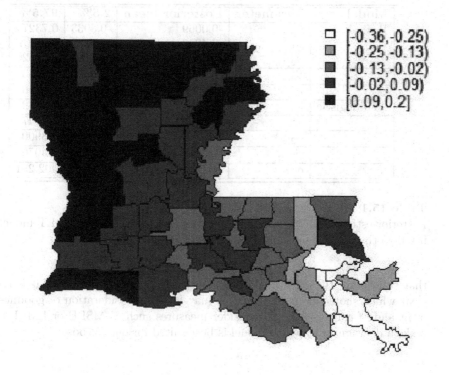

FIGURE 15.4

Posterior mean combined spatial effect (W) for the AFT model fitted to the Louisiana SEER data example.

Figure 15.4 displays the posterior mean of the combined spatial effect (W) for the AFT model.

The different models provide for some slightly differing interpretation (see Table 15.1). First neither model supports a well estimated race difference, in that neither of the levels of the race effect are significant. This suggests that the slight racial effect noticed in the KM plot is not substantive. For the Weibull model the age effect (β_1) appears to demonstrate a slight negative association, although the posterior marginal density upper tail is close to zero. However in the AFT model this effect is not well estimated. For either model most parameters are in fact not well estimated. As a footnote it should be noted that for these data stage at diagnosis is a major contributor to differences on survival, although this has not been demonstrated here in this example. Finally it should be clear that fitting different likelihood models is not commonly undertaken, and it is more usual to choose one a priori. From

Model	Parameter	Posterior mean	2.5%	97.5%
Weibull	β_0	-0.0069	-0.8033	0.7327
	β_1	-0.4246	-1.7252	-0.0161
	$\beta_{2(1)}$	0.00241	-0.7708	0.6752
	$\beta_{2(2)}$	-0.1379	-1.2680	0.5963
	v	2.2892	0.0803	7.1156
AFT logistic	β_0	-3.5551	-59.251	59.389
	β_1	-0.0176	-1.3410	1.1147
	$\beta_{2(1)}$	0.0166	-1.3066	1.3900
	$\beta_{2(2)}$	0.0122	-1.3620	1.3076
	σ	102.387	3.0410	402.231

Table 15.1
Posterior estimates of the main parameters for the Weibull and AFT models fitted to the SEER Louisiana PrCa data example.

that standpoint the AFT model has many advantages. However, if it is not clear which should be used then this may lead to consideration of goodness of fit and of course we could consider measures such as MSPE or LMPL to make a decision as to which model is best suited for our purpose.

16

Missingness, Measurement Error, and Variable Selection

In this chapter, the related topics of missing data, measurement error (ME), and variable selection will be reviewed. The connection between these topics may seem slight at first, but in fact there are subtle links. Missingness concerns the failure to observe a variable, whereas ME concerns partial observation. Variable selection often interacts with these effects and ultimately structural equation models (SEMs: (Stern and Jeon, 2004)) are built around partially observed groups of variables that are thought to affect an outcome or outcomes.

16.1 Missing Data

Missing data can arise often in biomedical studies and although it is less frequent when studying population level spatial health outcomes, it is prevalent when dealing with individual level patient data. There are different potential reasons for missingness and associated types of missingness. For purposes here random missingness (missing at random MAR; missing completely at random MCR) will be considered (Little and Rubin, 2019). Non random missingness (MNAR) can be accommodated by modeling the non- randomness mechanism directly. This will not be pursued here. For Bayesian models there are two different scenarios where missingness could arise and which demand different solutions. First is missingness in outcome variables, and the second is missingness in predictors.

16.1.1 Outcome Missingness

For outcome missingness, the posterior predictive distribution can be used with a data augmentation algorithm (Tanner, 1996). Define y_i^{obs}, $i = 1, ..., m$ as the observed outcome data, and y_j^{mis}, $j = m + 1, ..., m + n$ is the set of n missing outcomes. If we consider the missing outcomes as parameters we can first consider the posterior distribution as if we observed all the data:

$$P(\theta | \{y_i^{obs}\}, \{y_j^{mis}\}).$$

This is the conditional density of the parameters given the complete data. When sampled over the $\{y_j^{mis}\}$ this will approximate the density $P(\theta|\{y_i^{obs}\})$. Then in a second step we can sample $\{y_j^{mis}\}$ from the posterior predictive distribution:

$$P(\{y_j^{mis}\}|\{y_i^{obs}\}, \theta).$$

Cycles of sampling between these two distributions will provide a Bayesian multiple imputation of the missing outcomes. This augmentation algorithm is programmed automatically within BUGS and Nimble. All that is required is that missing outcome observations are denoted NA and then they are regarded as parameters to be estimated via augmentation. The following BUGS model code displays the missingness augmentation of a simple Weibull survival model with right censoring. In this example items 3 and 4 are censored at the time value 6.0 and so their end-points are not observed. Essentially the true end-point is missing and will be imputed. The times censored are listed as NA, whereas the actually censoring time is listed in tcen[]. Note that because of the missingness in the observed times they must have the missing values initialized in the inits.

```
model{
for(i in 1: 10){
t[i]~dweib(nu,lambda[i])C(tcen[i],)
lambda[i]<-exp(beta0+v[i])
v[i]~dnorm(0,tauv)
}
beta0~dnorm(0,0.0001)
nu~dgamma(1.0,0.0001)
tauv~dgamma(2,0.5)
}
##data
list(t=c(2.3,2.3,NA,NA,3.5,4.1,4.5,4.5,5.4,5.6),
tcen=c(0,0,6.0,6.0,0,0,0,0,0,0))
##inits
list(v=c(0,0,0,0,0,0,0,0,0,0),
t=c(NA,NA,7.0,7.0,NA,NA,NA,NA,NA,NA),
tauv=0.1,nu=0.1,beta0=0.1)
```

A single chain, run for 10,000 iterations with burnin 8000, yielded the estimates in Figure 16.1.

The initial values specified are both 7.0 and the results of a converged run show the estimated posterior mean values of 7.272 and 7.239. Although this is a slightly more complex missingness example as it involves censored data, it does serve to highlight the way that McMC packages can handle outcome missingness.

	mean	sd	MC_error	val2.5pc	median	val97.5pc	start	sample
t[3]	7.272	1.385	0.05924	6.035	6.85	11.12	8001	2000
t[4]	7.239	1.299	0.05659	6.028	6.853	10.7	8001	2000

FIGURE 16.1
Posterior predictive mean estimates of the missing times in item 3 and 4.

16.1.2   Predictor Missingness

Often models that include predictors assume that the predictors are constant, and the model conditions on the fixed predictor values. This is a standard assumption of much of statistical model theory. However if missingness occurs in predictors, then a mechanism must be considered by which the missing observations can be accommodated. *Complete case analysis (CCA)* is the solution whereby any items displaying missingness (whether outcomes or predictors) are removed from the data. Hence this is not an accommodation but an avoidance strategy. CCA is not regarded as a panacea for biomedical studies as it throws data away, and if intention to treat (ITT) is assumed, then CCA cannot be used.

A solution to the issue of fixed predictors is to assume that the predictors are a realization from a prior distribution. With a prior distribution assumed, any missing predictor values can be imputed as parameters. For continuous predictors it is often assumed that the true value is centered on an assumed prior mean with a fixed precision. In this way it is natural to assume a Gaussian prior distribution: $x_i \sim N(\mu_x, \tau)$ say. μ_x could be estimated from the sample mean of the observed predictor, and fixed. Also μ_x could have an assumed prior distribution and hence be allowed to be estimated also. Alternatively, a zero mean specification could be assumed. This would allow the missing predictor more freedom. The choice of precision however could be a critical parameter which could determine how far the imputed values would be allowed to deviate from the observed items. An estimate from the observed data based on the empirical variance (κ) could be used (e.g., $1/\hat{\kappa}$). This would mean that a data-based prior specification is assumed. Further, the precision could have a prior distribution assumed.

For discrete missing predictors, an appropriate categorical prior distribution could be assumed. In the binary case a Bernoulli distribution, for counts a Poisson or binomial distribution, and for multi-level categorical predictors a multinomial or singular multinomial could be used.

The following simulated example demonstrates the use of Nimble for illustration of the imputation of missing continuous predictors. The following R code generates a single predictor (x) and a Gaussian outcome (y). Items 5 and 7 are replaced by NAs in the predictor. The associated Nimble model is given below:

```
x<-rnorm(10,1,1)
xmean<-mean(x)
```

	Mean	Median	St.Dev.
x[5]	0.706904	0.677451	3.231691
x[7]	0.786917	0.704489	3.217597

FIGURE 16.2
Posterior mean estimates of the missing predictors for item 5 and 7.

```
mu<-1.0+2.0*x
mx<-rep(0.816,10)
y<-rnorm(100,mu,1)
x[5]<-NA
x[7]<-NA
library(nimble)
MOD<-nimbleCode({
for (i in 1:10){
y[i]~dnorm(mu[i],tauy)
mu[i]<-b0+b1*x[i]
x[i]~dnorm(mx[i],taux)
}
b0~dnorm(0,1000)
b1~dnorm(0,1000)
tauy~dgamma(2,0.5)
taux~dgamma(2,0.5)
})
```

The prior mean for the x predictor is assumed to be constant (0.816) and the regression parameters are assumed to have zero mean Gaussian prior distributions with fixed precisions (1000).

```
MODdata<-list(y,x)
names(MODdata)<-c("y","x")
MODCON<-list(mx)
names(MODCON)<-"mx"
MODinits<-list(tauy=0.1,taux=0.1,b0=-10,b1=0.1,x=c(NA,NA,NA,NA,0.816,
NA,0.816,NA,NA,NA))
samples<-nimbleMCMC(code=MOD,inits=MODinits,data=MODdata,
constants=MODCON,niter=2000,nburnin=1000,nchains=1,summary=TRUE)
```

With the above code the resulting summary produced x[5] and x[7] imputations in Figure 16.2.

Other times in the x vector have zero variation as they are fixed observations.

16.2 Measurement Error (ME)

Measurement error can be considered as a form of missingness where variables are partially observed (as opposed to completely missing). Here I provide a brief overview of basic ME ideas. For an in-depth examination of ME issues see Gustafson (2004), or Carroll et al. (2006). There are different approaches to ME depending on whether the variable is discrete or continuous. Discrete ME arises when a discrete variable is misrecorded. This could lead to misclassification error and is commonly found in social survey data, such as questionnaire responses (see Lesaffre and Lawson, 2012, Chapter 13 for more information). Here I focus on continuous ME where a measurement is not correctly observed but is on a continuous scale. There are two basic approaches to continuous ME: Berkson ME and classical ME. Assume the error occurs in predictors in a model: $\mathbf{x}_i$. In Berkson ME the true value of the misrecorded variable is assumed to be $x_i^T = x_i + \epsilon_i$ and this is substituted into the outcome model. For a Gaussian outcome model this could be $y_i \sim N(\mu_i, \tau^{-1})$ with $\mu_i = \beta_0 + \beta_1 x_i^T = \beta_0 + \beta_1(x_i + \epsilon_i)$ and τ is the precision. In this situation then ϵ_i is a random effect (a latent variable in fact). Often a zero mean Gaussian prior distribution would be assumed for this effect : i.e., $\epsilon_i \sim N(0, \tau_\epsilon^{-1})$. In the classical ME approach it is assumed that $x_i = x_i^T + \epsilon_i$ and as we assume that $y_i \sim N(\mu_i, \tau^{-1})$ with $\mu_i = \beta_0 + \beta_1 x_i^T$ then a joint model must be assumed for the estimation of x_i^T :

$$
\begin{aligned}
y_i &\sim N(\mu_i, \tau^{-1}) \\
\mu_i &= \beta_0 + \beta_1 x_i^T \\
x_i &= x_i^T + \epsilon_i \\
\epsilon_i &\sim N(0, \tau_\epsilon^{-1})
\end{aligned}
\tag{16.1}
$$

Suitable prior distributions for the regression parameters would also be of the form $N(0, \tau_\beta^{-1})$.

16.2.1 A Classical ME Example

A birth defect is recorded among live births and the geocode is residential address. Hence a binary outcome (birth defect or no defect) is available at a point location. An environmental insult consisting of heavy metal arsenic measured in top soil is available in a network of sites. Available interpolated data was limited to the residential locations of the individuals. The estimated arsenic level was obtained using universal Kriging. However no estimation error was available. Hence, it is reasonable to assume a continuous measurement error model for the arsenic in this example.

The code below shows an example of a logistic regression where a defect at birth is a binary outcome variable (defect: 0/1) and is linked to a linear predictor including birth-weight, baby gender, and measured soil concentration of arsenic. It is assumed that the birth defect response would be related to the true arsenic level in the soil, and so the variable included is AsT[i]. As[i] is the measured arsenic and is assumed to contain error. A continuous error model seems appropriate for the soil chemical metal and so the classical measurement error model is assumed whereby the disease outcome is modeled jointly with the soil chemical which has error. In this case, the defect outcome is defined to have $y_i \sim Bern(p_i)$ with a logit link to a set of predictors (including the true arsenic). The arsenic is assume to have a prior model of the form $as_i \sim N(as_i^T, \tau)$.

The logit model is formed from the different predictors as $\text{logit}(p_i) = a_0 + v_i + a_1 we_i + a_2 bs_i + a_3 as_i^T$. where we_i and bs_i are baby weight and baby sex respectively. As the true arsenic value is unobserved then it must also have a prior distribution. In this case AsT[i]~dnorm(As[i],10) so that the true value is constrained to lie close to the observed value. The precision is fixed in the example but could also have a hyperprior distribution. The BUGS code for this is given below. Nimble code is essentially identical in this case.

```
model{
for( i in 1:144){
defect[i]~dbern(p[i])
logit(p[i])<-a0+linP[i]+v[i]
As[i]~dnorm(AsT[i],tauAs)
linP[i]<-a[1]*(WEIGHT[i]-Wbar)+
a[2]*BABYSEX[i]+a[3]*(AsT[i]-Asbar)
v[i]~dnorm(0,tav)
AsT[i]~dnorm(As[i],10)
}
for (j in 1:3){
a[j]~dnorm(0,tava[j])
tava[j]~dgamma(2,0.5)
}
tauAs~dgamma(2,0.5)
tav~dgamma(2,0.5)
a0~dflat()
})
```

Because the true level of arsenic is a latent variable it must be initialized as a parameter vector. The data is as follows:

```
#DATA
list(Asbar=2.679,Wbar=3153.2,defect = c(0, 0,
0, 1, 0, 0, 0, 0, 1, 1, 0, 1, 0, 0, 0, 1, 0, 1, 0, 1, 1,
0, 0, 1, 0, 0, 1, 0, 1, 0, 0, 1, 0, 1, 0, 0, 0, 1, 0, 1,
```

0, 1, 1, 0, 0, 0, 1, 0, 1, 1, 0, 0, 1, 1, 1, 0, 0, 0, 0,
1, 0, 0, 0, 0, 0, 0, 0, 0, 0, 0, 0, 1, 0, 1, 1, 0, 0, 0, 0,
0, 0, 0, 0, 0, 0, 0, 0, 0, 0, 0, 0, 0, 0, 0, 0, 0, 1, 0, 0,
0, 0, 1, 0, 1, 0, 0, 1, 0, 0, 0, 1, 0, 0, 0, 0, 0, 0, 1, 0,
0, 1, 1, 0, 0, 0, 0, 0, 0, 0, 1, 0, 0, 0, 0, 0, 0, 0, 1, 0, 1,
0, 0, 0, 0, 0, 0, 0, 0, 0),
As = c(1.1607603463, 2.6140456724,
1.3567734145, 7.3826812556, 1.6723145295, 1.6311597846, 2.1911706214,
1.745341818, 3.7487806352, 7.3826812556, 2.2557737241, 2.5643437977,
2.6020754143, 2.3914767321, 3.0328633365, 0.9989822, 2.5371645629,
1.4696777656, 2.4272795768, 2.4272795768, 2.4314489945, 1.7947152887,
1.8560598618, 3.781001436, 0.5623485464, 2.5878403305, 2.666936828,
2.6568931692, 4.1965187823, 3.0029315067, 2.142537184, 2.5499496794,
1.8013390023, 3.8447587814, 4.2383600904, 2.9789560257, 2.4136248265,
2.3754678842, 2.369883201, 2.6684642235, 3.0351120132, 1.8844644155,
1.7367947287, 2.8120449157, 2.6684642235, 2.6468626963, 3.7593232592,
4.153909314, 2.7277989079, 3.6561589105, 2.015873234, 3.506763599,
0.4442933673, 7.3826812556, 3.3549823412, 1.168649723, 2.209088732,
2.015873234, 2.1899187102, 2.142537184, 2.4724171718, 2.595866711,
1.3578982243, 2.798647705, 2.5194294981, 1.4969806872, 3.5537099623,
1.9780802994, 4.2559514727, 2.8171322661, 4.1641713564, 1.3578982243,
1.9368800182, 7.3925383497, 4.2095911969, 2.4218399505, 1.1689041032,
3.9067577705, 2.2431274854, 2.4622511482, 2.5742791248, 2.139347526,
0.9784962066, 1.0989208688, 2.4280051222, 2.4144718019, 1.519834074,
1.0947733642, 1.7947152887, 0.7457245722, 1.2973253387, 2.561987819,
3.6181125928, 4.3099397612, 1.1014986099, 2.3912670338, 3.600088507,
2.3781846101, 2.3441117602, 7.3925383497, 3.9801899613, 3.5626492068,
7.3925383497, 2.4694564626, 4.3099397612, 2.55103649, 2.5574240984,
2.8917778994, 3.243591498, 2.5315418189, 2.4268939171, 1.7395911785,
0.7791685181, 1.7749323977, 2.234713335, 2.7857601474, 1.1394211466,
2.9286641682, 2.6034937371, 2.4341290089, 4.7557411394, 1.6663445364,
1.817066625, 2.2363666644, 2.2819441527, 3.8815448612, 2.5638401184,
3.1100617875, 1.8976972816, 2.2985952182, 4.6081126454, 1.649773575,
2.4648513037, 2.143839497, 1.9770018174, 2.5583664138, 7.3925383497,
2.5435946928, 1.9232480908, 2.3896606632, 1.7285569097, 1.155739397,
2.4095968014, 1.1014986099),
WEIGHT = c(2807,
3833, 2309, 2802, 2917, 3270, 3201, 3011, 3761, 1622, 3223, 2435,
3128, 2460, 1142, 3272, 3146, 3902, 3721, 2354, 3213, 3290, 1809,
2540, 3254, 3027, 3020, 3118, 2775, 3612, 3040, 3784, 3485, 2570,
3287, 3523, 3052, 3585, 3658, 4120, 2850, 2770, 3213, 3036, 2600,
2302, 3346, 3164, 2716, 3325, 3562, 3710, 3867, 3691, 2965, 3447,
3719, 2995, 3504, 2697, 3473, 3104, 3312, 3817, 3027, 3023, 2665,

	mean	sd	2.50%	97.50%
a[1]	-1.99E-04	4.64E-04	-0.0012	7.08E-04
a[2]	-0.8625	0.5915	-2.279	0.128
a[3]	0.5217	0.2016	0.1533	0.9482
a0	-0.1807	0.8301	-1.774	1.534

FIGURE 16.3

Posterior regression parameter estimates for the soil arsenic example

```
2817, 2973, 2714, 3391, 2932, 2785, 2890, 3296, 3082, 3287, 3166,
2380, 4255, 3292, 3468, 3284, 3148, 3346, 3122, 3617, 3637, 3813,
4350, 3503, 3395, 3426, 2780, 3939, 2230, 3244, 3043, 4108, 993,
3684, 3455, 3105, 3120, 3439, 3012, 3350, 4003, 2806, 3087, 3324,
3157, 3233, 3147, 2764, 3604, 3322, 2601, 3852, 3273, 3267, 2994,
2907, 3498, 2854, 3545, 3912, 3514, 3620, 2211, 2920, 4026, 3062,
2098, 3430, 3430, 2722, 3430, 2381, 3374, 2695, 3487, 2438, 3232
),BABYSEX = c(1, 2, 2, 1, 2, 1, 1, 1, 1, 2, 2, 1, 1, 2, 1,
1, 2, 1, 1, 1, 1, 2, 2, 1, 2, 2, 1, 2, 2, 2, 1, 1, 1, 1, 2, 1,
2, 1, 1, 1, 2, 2, 1, 1, 2, 2, 1, 1, 2, 1, 2, 2, 1, 2, 2, 1, 1,
2, 1, 2, 1, 2, 2, 2, 2, 1, 2, 1, 1, 1, 1, 2, 1, 2, 1, 2, 2, 1,
2, 1, 2, 1, 2, 2, 2, 2, 1, 1, 1, 1, 1, 1, 1, 1, 1, 1, 1, 1, 1,
1, 2, 1, 2, 1, 1, 2, 1, 1, 2, 1, 1, 2, 2, 2, 2, 1, 1, 1, 1, 1,
2, 2, 1, 1, 2, 1, 1, 1, 1, 1, 2, 1, 2, 2, 1, 2, 1, 2, 2, 1, 2,
1, 2, 1))
```

Figure 16.3 displays the BUGS version output of the above code from a converged sampler (after 8000 burnin). In this case both weight and baby sex estimates cross zero and they are not well estimated. The arsenic parameter is positive and is well estimated (CI: 0.1533, 0.9482), which suggests that there is evidence for an association between the measured arsenic and the birth defect outcome.

The resulting posterior mean estimates for true arsenic (AsT) for this model are given in Figure 16.4:

Note that these are close to the observed values; but as the precision on the true arsenic prior distribution is reduced then the further the true arsenic will deviate from the observed values.

	mean	2.50%	97.50%
AsT[1]	1.161	1.106	1.211
AsT[2]	2.614	2.558	2.666
AsT[3]	1.357	1.306	1.409
AsT[4]	7.383	7.33	7.434
AsT[5]	1.673	1.622	1.726
AsT[6]	1.631	1.579	1.687
AsT[7]	2.191	2.136	2.245
AsT[8]	1.746	1.697	1.8

FIGURE 16.4
Posterior mean estimates of AsT for the soil arsenic example.

16.3 Variable Selection

In multiple regression examples it is often the case that some form of variable selection (VS) is appropriate. This is true for GLMs and GLMMs and BHMs in general. This could be required to reduce the dimensionality of the model, or simply to assess which predictors are important in explaining the variation in the outcome. For Bayesian models there are a number of approaches, ranging from model selection to penalized lasso-type dimension reduction. Here I will focus on stochastic search algorithms for VS (SSVS). An excellent review paper is O'Hara and Sillanpää (2009), and more background can be found in Dellaportas et al. (2002), or Petralias and Dellaportas (2013). The four main methods of SSVS are Gibbs VS (GVS), spike and slab VS, Laplace shrinkage, and reversible jump McMC. Here I will demonstrate two of these approaches: GVS and Laplace shrinkage.

16.3.1 Gibbs VS

In the case of Gibbs variable selection we will demonstrate the use of the Kuo and Mallick (Kuo and Mallick (1998)) entry parameter approach. In this case we hypothesize that the j th variable can be sampled and its indicator of inclusion (entry) is I_j. The prior joint probability of the regression parameter and entry parameter is defined by $\Pr(\beta_j|I_j) = \Pr(\beta_j)\Pr(I_j)$. We can assume a Bernoulli prior distribution for I_j as it reflects the inclusion or otherwise of the variable. β_j would usually have a zero mean Gaussian prior distribution. Hence, for each parameter we would have $\Pr(\beta_j|I_j) = N(0,\tau^{-1}).Bern(p_j)$. Within the model with a linear predictor including G predictors we assume

$$\eta_i = \beta_0 + \sum_{j=1}^{G} I_j\beta_j x_{ji},$$

so that when the indicator is sampled the predictor is either included in the model or excluded. Code for this type of Gibbs VS is given below. The example is for Georgia counties and low birth weight counts in the year 2007. The total birth count is also available and it can be assumed that a spatial logistic model is appropriate: i.e.,

$$y_i \sim bin(p_i, n_i)$$
$$\log it(p_i) = \beta_0 + \sum_{j=1}^{G} I_j\beta_j x_{ji} + v_i.$$
$$v_i \sim N(0,\tau_v^{-1})$$

Here it is assumed that the random effect component has a zero mean Gaussian prior distribution and hence is uncorrelated a priori. There are five predictors in the model: population density, black proportion, median income, % under the poverty line, unemployment rate ($x1[]....x5[]$).

```
nimbleCode({
for(i in 1:159){
x1c[i]<-(x1[i]-mean(x1[1:159]))/sd(x1[1:159])
x2c[i]<-(x2[i]-mean(x2[1:159]))/sd(x2[1:159])
x3c[i]<-(x3[i]-mean(x3[1:159]))/sd(x3[1:159])
x4c[i]<-(x4[i]-mean(x4[1:159]))/sd(x4[1:159])
x5c[i]<-(x5[i]-mean(x5[1:159]))/sd(x5[1:159])
y1[i]~dbin(p1[i],n[i])
logit(p1[i])<-b0+b[1]*psi[1]*x1c[i]+b[2]*psi[2]*x2c[i]+b[3]*psi[3]*x3c[i]+b[4]*
psi[4]*x4c[i]+b[5]*psi[5]*x5c[i]+v[i]
v[i]~dnorm(0,tauV)
}
for( j in 1: 5){
psi[j]~dbern(p[j])
p[j]~dbeta(0.5,0.5)}
tauV<-pow(sdV,-2)
sdV~dunif(0,10)
b0~dnorm(0,taub0)
for(j in 1:5){b[j]~dnorm(0,taub[j])
taub[j]<-pow(sdb[j],-2)
```

Variable	I_j	β estimate (sd)
1	0.0276	-0.01368 (1.116)
2	1.0	0.1129 (0.021)
3	0.0798	0.0161 (1.131)
4	0.9992	0.1147 (0.023)
5	0.06663	0.0087 (1.100)

Table 16.1
Converged results for the 5 parameter low birth weight binomial model

```
sdb[j]~dunif(0,10)
}
taub0<-pow(sdb0,-2)
sdb0~dunif(0,10)
})
```

In this example, the Bernoulli probability is given a beta (Jeffrey's) prior distribution ($beta(0.5, 0.5)$), which allows the probability to be estimated during sampling. The posterior average of psi[] yields the inclusion or entry probability for a given predictor. As a threshold for acceptance of a predictor, it is often assumed that a value of $ave(I_j) > 0.5$ is relevant. This was found to be optimal for Gaussian models (Barbieri and Berger, 2004) and has been adopted as a general criterion. For the low birth weight example, the following table displays the converged results for the posterior mean ($ave(I_j)$) and associated regression parameter estimates with standard deviations. It is clear in this case that variable 2 (% black) and 4 (% under poverty line) pass the inclusion threshold while the others do not. The associated regression parameters are well estimated for these predictors. Hence the choice of a final model with only predictors 2 and 4 would be a reasonable decision.

Note however that collinearity in predictors could still lead to values of $ave(I_j)$ less than 0.5 due to the fact that collinear predictors could be swapped in models.

16.3.2 Laplace Shrinkage

Laplace shrinkage is a simpler approach to selection in that it assumes a penalization of the prior distribution for the regression parameters. In essence a Gaussian prior distribution is replaced by double exponential distribution which has an asymptote at zero. This puts greater prior weight on removal of the predictor(s). The following is nimble code for the shrinkage example:

```
nimbleCode({
 for(i in 1:159){
 x1c[i]<-(x1[i]-mean(x1[1:159]))/sd(x1[1:159])
```

Variable	β estimate (sd)
1	0.00084 (0.0129)
2	0.1127 (0.0259)
3	0.0133 (0.0273)
4	0.1308 (0.0315)
5	-0.0062 (0.025)

Table 16.2
Converged results for the 5 parameter low birth weight binomial model:
Laplace shrinkage

```
x2c[i]<-(x2[i]-mean(x2[1:159]))/sd(x2[1:159])
x3c[i]<-(x3[i]-mean(x3[1:159]))/sd(x3[1:159])
x4c[i]<-(x4[i]-mean(x4[1:159]))/sd(x4[1:159])
x5c[i]<-(x5[i]-mean(x5[1:159]))/sd(x5[1:159])
y1[i]~dbin(p1[i],n[i])
logit(p1[i])<-b0+b[1]*x1c[i]+b[2]*x2c[i]+b[3]*x3c[i]+b[4]*x4c[i]+b[5]*
x5c[i]+v[i]
v[i]~dnorm(0,tauV)
}
tauV<-pow(sdV,-2)
sdV~dunif(0,2)
b0~dnorm(0,taub0)
for(j in 1:5){b[j]~ddexp(0,taub[j])
taub[j]~dgamma(2,1)
}
taub0~dgamma(2,1)
})
```

Table 16.2 displays the converged results of fitting the Laplace shrinkage model to the LBW Georgia example. As for the GVS approach the well estimated predictors are (2) and (4) , whereas the others have 95% credible intervals crossing zero (not shown).

In this example it is clear which variables are making a major contribution. However there are many cases where such a clear picture does not emerge. The problem of collinearity could lead to increase in standard errors for parameters and, in the GVS case, many inclusion probabilities just less than 0.5. In these cases there may be need for more sophisticated approaches such as partial least squares or multi stage dimension reduction to better model the relations (Rockova and George, 2014).

Alternative purpose-built R libraries are available for fitting variable selection models via McMC. One example is spikeSlabGAM which provides for SSVS with a range of data models and also non-linear link functions to predictors with addition of spatial random effects.

17

Individual Event Modeling

17.1   Specific Spatial Modeling

By specific it is meant that a locational geocode is the focus of the individual health outcome analysis. In this sense this is a directly spatial analysis (rather than a spatial contextual analysis). For example, in environmental epidemiology it is a common theme that proximity to potential insults (e.g., sources of pollution) should be a primary concern in the assessment of disease risk. This can take two forms. First location of an individual (such as residence) is related to the 'local' environment at or near that location. In this case measures of potential insults could be evaluated at or near the location. These could be interpolated to the location (of residence). A classic example would be particulate matter in the air (PM2.5 or 10) could be thought to relate to asthma incidence. If residential location of asthma sufferers is available then PM2.5 or PM10 measured at monitor sites in the vicinity of the residences could be interpolated to location of residence (Kibria et al., 2002, Chang, 2016). In this case, the disease outcome is y_i at the i th location (coordinates: $s_i : (x_i, y_i)$, $i = 1, ..., m$). Assume that we have a set of n measurement sites for an environmental insult (z_j, $j = 1, ..., n$). A model is required for the interpolation to the m individual locations. Ultimately a joint model for the disease outcome and interpolation must be assumed.

For example,

$$y_i \sim Bern(p_{s_i})$$
$$\text{logit}(p_{s_i}) = \alpha_0 + f(z_i^*) + R_i$$
$$[z_i^* | \{z_j\}, \boldsymbol{\theta}] \sim \mathbf{N}(\boldsymbol{\mu}, \boldsymbol{\Sigma})$$

where $f(z_i^*)$ is a function of the interpolated insult, could be posited. If fitted jointly, especially with McMC, the posterior sampling of z_i^* should correctly assign the uncertainty to the estimation of $f(z_i^*)$ within the health outcome model. The interpolated insults are generated from the predictive distribution $\mathbf{N}(\boldsymbol{\mu}, \boldsymbol{\Sigma})$ where $\boldsymbol{\mu}$, and $\boldsymbol{\Sigma}$ are constructed from $\boldsymbol{\mu} = \mathbf{x}^* \beta + C_{12}^T C_{11}^{-1}(\mathbf{z} - \mathbf{x}\beta)$ and $\boldsymbol{\Sigma} = \mathbf{C}_{12} - C_{12}^T C_{11} C_{12}$ where $\mathbf{x}$ and $\mathbf{x}^*$ is a vector of explanatory variables and C_{11} is the covariance matrix of $\mathbf{z}$, C_{12} is the cross covariance between $\mathbf{z}$ and $\mathbf{z}^*$ (Kim et al., 2010, Banerjee et al., 2014).

One disadvantage of this Bayesian Kriging approach is that joint modeling could lead to the estimation of $f(z_i^*)$ being influenced by the health outcome model (which appears counter-intuitive!). A number of approaches to circumvent this problem have been proposed (Dominici et al., 2003, Keller et al., 2017). A simple alternative is to consider a form of "plug-in" estimate where $\mathbf{z}^*$ is estimated separately ($\widehat{\mathbf{z}}^*$ say) so that $\text{logit}(p_{s_i}) = \alpha_0 + f(\widehat{z}_i^*) + R_i$ is estimated using the plug-in. This ignores the variation in the estimation of $\mathbf{z}^*$. An extension of this idea is to consider that the estimate is with error so that some allowance for the uncertainty is available. One proposal is to consider this as a measurement error problem . One approach is to consider Berkson error and to add a random component to $\widehat{\mathbf{z}}^*$ within the health outcome model. This can be achieved by scaling a standard error for the estimate of $\widehat{\mathbf{z}}^*$. Hence it is possible to use $\text{logit}(p_{s_i}) = \alpha_0 + f(\widehat{z}_i^* + \kappa\widehat{\sigma}) + R_i$ where $\widehat{\sigma}$ is the estimated standard error of $\widehat{\mathbf{z}}^*$ and the scaling κ has a suitable prior distribution (see e.g., Onicescu et al., 2014). Code for this in nimble, with $f(.) = \beta(.)$, with a single interpolated predictor could be as follows:

```
nimbleCode(
{for (i in 1:m){
y[i]~dbern(p[i])
logit(p[i])<-a0+beta*(zstar[i]+kappa*sez[i])+R[i]
#  R[i] can be a set of spatially referenced or individual random effects
}
kappa~dnorm (0,tauk)
beta~dnorm(0,taub)
a0~dnorm(0,taua0)
tauk~dgamma(2,0.5)
taub~dgamma(2,0.5)
taua0~dgamma(2,0.5)
})
```

In a related example, examined in Chapter 16, the environmental insult is heavy metal arsenic measured in top soil in a network of sites. Available interpolated data was limited to the residential locations of the individuals. The estimated arsenic level was obtained using universal Kriging. However no estimation error was available. Hence it is reasonable to assume a continuous measurement error model for the arsenic in this example. This is discussed in more detail in section 16.2.

Second, often it is assumed that proximity to an insult is relevant and the spatial association is defined by distance and/or direction from the source location. This is the classic putative source analysis found in many environmental health case studies (Lawson and Cressie, 2000, Elliott et al., 2007, Nieuwenhuisen, 2016). Usually the association between an insult in this case is via exposure surrogates such as distance from source, direction from source and interaction between these variables. In general is the outcome to be modeled is a discrete and a retrospective analysis is to be carried out, then we can define the outcome at m locations $(s_i : (x_i, y_i),\ i = 1, ..., m)$ as y_i. Also observed is the location of the insult. Usually this is a single location $s_c : x_c, y_c$. However there could also be multiple locations and the insult may arise not from an isolated point location. For example, road systems could be responsible for air pollution in the vicinity of a residence. The nearest road could be important, but may not be the only road with a potential contribution. Hence competing risks could be important. Here it will consider only a single putative source.

The distance and direction from the source are defined as $d_i = \sqrt{(x_i - x_c)^2 + (y_i - y_c)^2}$ and $\phi_i = \tan^{-1}(\frac{(y_i - y_c)}{(x_i - x_c)})$. Basic models of risk use these two components of polar coordinates. Many examples of distance-only modeling can be found.

A classic relative risk model could be for a binary outcome at locations

$$y_i \sim Bern(p_i)$$
$$\text{logit}(p_i) = \alpha_0 + f_1(d_i, \phi_i) + f_2(\mathbf{x}_i) + R_i$$

where $f_2(\mathbf{x}_i)$ is a function of covariates (possibly individual level), and R_i is a set of random effects (both individual and location specific). In the example examined here the location of $m = 58$ larynx cancer cases and $n = 978$ respiratory cancer cases in an area of North West England (Lancashire) are available. This dataset was first analyzed by Diggle (1990) and later by Diggle and Rowlingson (1994). Figure 17.1 displays the distribution of the two diseases and the location of a commercial incinerator which was thought to be a potential source of additional cancer risk. In this example the locations of the two diseases can be superimposed and a new binary variable constructed, whereby

$$y_i = \begin{cases} 1 & \text{larynx cancer} \\ 0 & \text{respiratory cancer} \end{cases}$$

Then $y_i \sim Bern(p_i)$ and the logit of the probability of larynx versus respiratory cancer (conditioning on location) can be modeled with a linear or non-linear predictor including personal covariates, locational effects, and random effects. For example

$$\text{logit}(p_i) = \gamma_0 + \gamma_1 d_i + \gamma_2 x_i + R_i \tag{17.1}$$
$$R_i = v_i + W_i$$

where x_i is the age of each person, d_i is the distance from the incinerator, and R_i could consist of two random components such as uncorrelated and spatially correlated effect or a composite geostatistical spatial effect. In the following nimble has been used to fit a convolution model for R_i, with a zero mean Gaussian prior distribution for v_i and a ICAR prior distribution for W_i, (M1) and a multivariate normal model with spatial exponential covariance for W_i (M2). In addition, CARBayes was used to fit the BYM model for R_i (M3). To

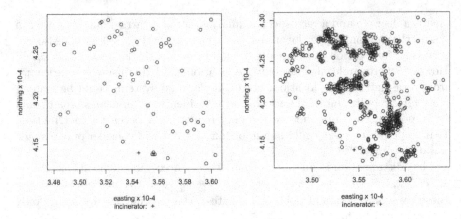

FIGURE 17.1
Larynx cancer case residential addresses (left) and respiratory cancer case addresses (right) in an area of North West England 1973–1984.

allow the use of conditional autoregressive prior distributions within nimble or CARBayes for individual location data, it is necessary to derive neighborhoods, and to either use adjacency information in BUGS format (adj, num format) or using binary matrix form (CARBayes).   Natural neighborhoods can be derived via tessellation or triangulation of the point locations. The Dirichlet tessellation provides a tiling of points whose adjacent edges indicate natural neighbors (Sibson, 1980, Okabe et al., 1992). The dual Delauney triangulation can also be used to define adjacencies. The R packages deldir and spatstat can be used to obtain this information. Figure 17.2 displays the tessellation of the larynx cancer example, which is used to derive neighbors. The following code can be used to obtain the relevant adjacency information.

```
#assume coordinates in vectors x and y
#assume a bounding rectangle.
#Here a rectangle slightly bigger than the maximum and minimum dimen-
sion is assumed
x<-jitter(x);y<-jitter(y)  ## making sure points are not coincident
library(spdep)
library(maptools)
library(spatstat)
maxx<-max(x)+0.01;minx<-min(x)-0.01;maxy<-max(y)+0.01;miny<-
min(y)-0.01
Xmat<-cbind(x,y)
X<-as.ppp(Xmat,W=c(minx,maxx,miny,maxy))
Z<-dirichlet(X)
```

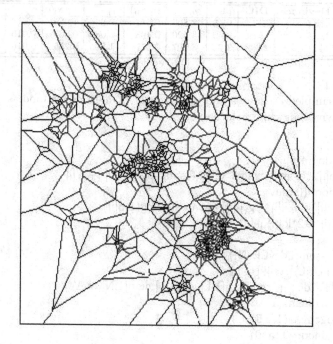

FIGURE 17.2
Dirichlet tessellation of the larynx and respiratory cancer case event example.

```
spd<-as(Z, "SpatialPolygons") ## Z is a tess object and spd is a spatial
polygon object
Larnb<-poly2nb(spd)
####   for CARBayes binary matrix ###
W.mat<-nb2mat(Larnb,style="B")
#### for adj and num vectors #######
adjnum<-nb2WB(Larnb)
adj<-adjnum$adj
num<-adjnum$num
```

The code for the ICAR convolution model (M1) is below:

```
IndCAR<-nimbleCode(
{
for (i in 1:N){
```

Model	Deviance	DIC	pD	γ_0	sd(γ_0)	γ_1	sd(γ_1)	γ_2	sd(γ_3)
M1	418.9	514.3	95.5	-3.806	0.511	-0.606	1.969	0.013	0.007
M2	433.7	467.6	33.9	-3.520	0.463	-0.086	0.669	0.016	0.007
M3	444.71	448.4	3.7	-3.707	0.577	-1.406	3.755	0.016	0.008

Table 17.1
Posterior sampled estimates and goodness of fit metrics for models M1- M3
for the larynx-lung cancer data example.

```
ind[i]~dbern(p[i])
logit(p[i])<-gam0+gam1*dis[i]+gam2*age[i]+v[i]+W[i]
v[i]~dnorm(0,tauv)
log(Lik[i])<-ind[i]*log(p[i])+(1-ind[i])*log(1-p[i])
LdevC[i]<-2*log(Lik[i])
}
DevC<-sum(LdevC[1:N])
 for(k in 1:L){wei[k]<-1}
 W[1:N] ~dcar_normal(adj[1:L],wei[1:L],num[1:N],tauW)
tauW~dgamma(1,0.5)
tau0~dgamma(1,0.5)
gam0~dnorm(0,tau0)
gam1~dnorm(0,tau1)
gam2~dnorm(0,tau2)
tauv~dgamma(1,0.5)
tau1~dgamma(1,0.5)
tau2~dgamma(1,0.5)
})
```

The code for the nimble geostatistical model (M2) is below:

```
disSq<-matrix(0,nrow=N,ncol=N)
for (i in 1:N){
for (j in 1:N){
disSq[i,j]<-(y[i]-y[j])**2+(x[i]-x[j])**2
disSq[i,j]<-disSq[i,j]
}}
###########################
indlar<-nimbleCode({
for (i in 1:N){
ind[i]~dbern(p[i])
logit(p[i])<-gam0+gam1*dis[i]+gam2*age[i]+v[i]+Wcov[i]
v[i]~dnorm(0,tauv)
log(L[i])<-ind[i]*log(p[i])+(1-ind[i])*log(1-p[i])
LdevC[i]<-2*log(L[i])
```

```
}
cov[1:N,1:N]<-invTau*exp(-alp*disSq[1:N,1:N])
Wcov[1:N]~dmnorm(zero[1:N],cholesky =cov[1:N,1:N], prec_param=0)
DevC<-sum(LdevC[1:N])
invTau~dgamma(1,0.5)
alp~dgamma(1,0.5)
gam0~dnorm(0,tau0)
gam1~dnorm(0,tau1)
gam2~dnorm(0,tau2)
tauv~dgamma(2,0.5)
tau0~dgamma(2,0.5)
tau1~dgamma(2,0.5)
tau2~dgamma(2,0.5)
})
```

The code for the CARBayes convolution model (M1) is below:

```
data1<-data.frame(ind,dis,age,denom)
form<-ind~1+dis+age
res<-S.CARbym(form, family="binomial", data=data1, trials=denom,
W=W.mat, burnin=16000,
   n.sample=20000, thin=1,verbose=TRUE)
gam0M<-mean(res$samples$beta[,1]);gam0SD<-sd(res$samples$beta[,1])
gam1M<-mean(res$samples$beta[,2]);gam1SD<-sd(res$samples$beta[,2])
gam2M<-mean(res$samples$beta[,3]);gam2SD<-sd(res$samples$beta[,3])
```

Table 17.1 displays the results of fitting these models to the larynx-respiratory cancer dataset from NW England. The logistic model defined in 17.1 was fitted and posterior estimates of the regression parameters as well as goodness of fit diagnostics were obtained. The deviance and DIC were computed following a burnin of 16000 iterations with a single chain. Convergence was checked using Geweke diagnostics from the coda package. The deviances reported are similar in magnitude for each approach but the effective number of parameters differs quite markedly. Between M1 and M2 the geostatistical model has lower DIC, mainly due to the smaller number of parameters. The large difference between M1 and M3 in terms of resulting DIC is problematic, but could be related to the fact that the conservative estimate of the pD is used in the nimble examples. It is clear that, while each approach yields well estimated intercept and age coefficient, the distance effect (γ_1) is not well estimated. It is also notable that these γ_1 estimates vary greatly.

Note that binary models with logistic links to spatial effects could also be fitted using McMC with SPBayes (or formerly on older R versions on GeoRglm) and this would allow the direct estimation of spatial covariance effects. INLA can also be used (see Gomez-Rubio, 2020). I do not pursue this here. It should be noted that using distance alone in these putative source models can lead to interpretational problems as risk-distance relations are not always simple. In general, it is better to consider both distance and direction, especially when dealing with air and water pollution insults. Correct exposure modeling for putative source problems is in fact a major challenge, and incorrect assumptions can lead to erroneous public health decisions.

17.2   Individual and Aggregate Contextual Outcome Modeling

Individual contextual modeling is in fact just a special case of the application of BHM where spatial or spatio-temporal contextual effects are included. In general, assume an individual level model for the i th individual y_i with design vector of individual predictors: x_i^t $i = 1, ..., m$. Also assuming a generalized linear mixed model for the outcome, then the data model would be $y_i \sim g(x_i^t \beta + R_i)$. Let the contextual level be denoted by superscript number ($l = 1,2,3....$). The contextual model with an arbitrary number of levels could be $y_i \sim g(x_i^t \beta + \mathbf{x}_i^{*t} \gamma + \sum_l r_i^l)$ where the data model $g(.)$ could include a suitable link function. Here $x_i^t \beta$ is a linear predictor based on the individual level predictors, and $\mathbf{x}_i^{*t} \gamma$ is a predictor based on the individual contextual (inherited) predictors (so that an individual inherits a predictor if they are within a l th contextual group). A contextual group could be one of a variety of effects. For example, they could include family groups, association with attendance at a school, age x gender groups, ethnicity, and also spatial groupings such as zip/postal code, block-group, census tract, or county or municipality. Hence, x_i^t could include individual age, gender, ethnicity, income, education level, whereas $\mathbf{x}_i^{*t}$ could include, for example: x_i^a- age group factor, x_i^s- school factor , $x_i^{bg}-$ predictor measured at census block-group in which the person resides, (i.e., x_i^z where $i \in j$ and j denotes the j th block-group), $x_i^{ct}-$ census tract, $x_i^{co}-$ county. Hence a model with median income measured as x_i^{bg}, x_i^{ct}, and x_i^{co} would be truly hierarchical in that the individual would inherit the effects of different scales of neighborhood (block-group, census-tract, and county) which are all aggregate aligned. Note that the term $\sum_l r_i^l$ represents the sum of random effects at different contextual levels. In the next example, there is an individual (frailty) effect and then two county-level contextual effects.

The following example of code is for an individual binary outcome (low birth weight : LBW) for a set of individual births. The set of individual births is a subset of all South Carolina births for the year 2000. The first 100 births are used here for purpose of exposition. Each birth has associated a number of individual level predictors: mothers age, mothers race (3 levels), county label (in alphabetic order).

```
LBWcode<-nimbleCode({
for (i in 1 : 100){
LBW[i]~dbern(p[i])
logit(p[i])<-a0+a1*AGEM[i]+RACE[RACEM[i]]+V[i]+V1[Clabel[i]]+
W[Clabel[i]]
V[i]~dnorm(0,tauV)
}
for (j in 1:3){RACE[j]~dnorm(0,tauR)}
for (i in 1:L){wei[i]<-1}
for (j in 1:46){V1[j]~dnorm(0,tauV1)}
W[1:46]~dcar_normal(adj[1:L],wei[1:L],num[1:46],tauW,zero_mean=1)
tauV~dgamma(2,0.5)
tauW~dgamma(2,0.5)
tauR~dgamma(2,0.5)
a0~dnorm(0,tau0)
a1~dnorm(0,tau1)
tau0~dgamma(2,0.5)
tau1~dgamma(2,0.5)
})
```

Figure 17.3 displays the boxplots of the final sample (of size 4000) for the first 60 individual uncorrelated random frailty effects.

Figure 17.4 displays the resulting county level posterior mean uncorrelated and correlated effects for the fitted model with mother's age and mother's race as covariates, and an individual frailty effect ($v[i]$), and an uncorrelated county effect (V1) and a correlated ICAR effect also at county level (W).

There is clearly substantial spatial structure in these data judging by the clustered appearance of the correlated effect (W). The posterior mean estimates and SDs for the main regression parameters are shown in Table 17.2.

These results suggest that while the intercept is not well estimated, the age coefficient appears to be quite well estimated and the 95 % credible interval for this parameter presents a limited excursion across zero (not shown). The density estimate for the parameters demonstrates that the bulk of the probability mass seems to be on the negative side (Figure 17.5). This suggest that increasing age has a slight negative effect on birthweight on the logistic scale.

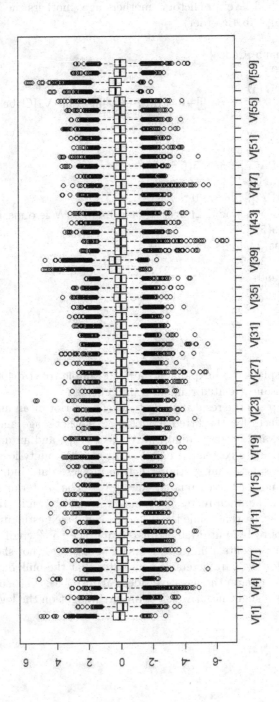

FIGURE 17.3
Boxplots of first 60 individual's uncorrelated REs in the final posterior sample of size 4000.

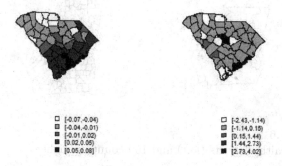

FIGURE 17.4
Posterior mean estimates of the correlated contextual county effect (W: left)
and the uncorrelated contextual county effect (V1: right) for the Bernoulli
low birth weight model with mother's age and mother's race as covariates.

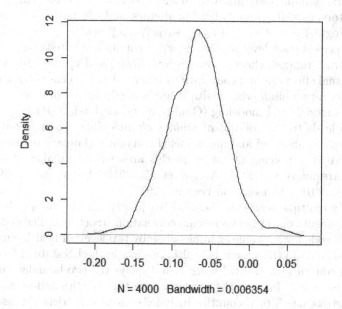

FIGURE 17.5
Density estimate plot of the final sample of parameter a1 from the fitted
logistic spatial model for the individual low birth weight example.

FIGURE 17.6
Georgia 18 health districts (left) and 159 counties (right).

Note that while the example above refers to individual level effects, it is straightforward to apply the same contextual rationale to aggregate count data. For example, if the census tract-level counts of low birth weight are observed as y_i^{ct} where i now denotes the i th census tract then, assuming a Poisson data model for these counts, with census tract expected rate e_i^{ct} i.e.,

$$y_i^{ct} \sim Pois(e_i^{ct}\theta_i^{ct})$$

then we could assume a log link with linear predictor based on census tract level predictors, as well as county level predictors, and effects. In this case, we could have $\log(\theta_i) = a_0 + a_1 x_{1i}^{ct} + a_2 x_{2i}^{ct} + a_3 x_{1i}^{co} \dots v_i^{ct} + u_i^{ct} + v_i^{co} + u_i^{co}$ where x_{1i}^{ct}, x_{2i}^{ct} are census tract level predictors, x_{1i}^{co} a county level predictor, v_i^{ct}, u_i^{ct} are census tract random effects (convolution effects) and v_i^{co}, u_i^{co} are county level contextual effects (also a convolution) inherited at the census tract level. This provides a way of allowing multiple levels of effects and in essence leads to multilevel models and modeling (Goldstein and Leyland, 2001).

This also leads to the concept of *multi-scale* modeling, whereby multiple levels of *data* are observed and are modeled conjointly (Louie and Kolaczyk, 2006). There is a growing literature in this area with a number of recent additions (Aregay et al., 2014, Aregay et al., 2015, Aregay et al., 2016b, Aregay et al., 2016a, Fonseca and Ferreira, 2017).

Essentially multiple scales can be modeled jointly with linkages either via shared components or via between-scale correlation structures. The code below is BUGS code for a simple example whereby two levels of health outcome data are observed. In this case it is health districts in the US state of Georgia (p=18) and counties (m=159). Figure 17.6 displays the two boundary maps. Note that the health districts align exactly with the counties and on average a health district has 5 or 6 counties included. Health districts are used to allocate resources for public health management.

```
# p=18 districts m=159 counties
model{
```

Parameter	Mean	SD
a0	-0.246	0.668
a1	-0.0689	0.0379
race[1]	-0.258	0.485
race[2]	-0.0372	0.470
race[3]	0.0886	0.557

Table 17.2
Posterior mean estimates and SDs for the 4000 size sample from the converged sampler for the individual low birth weight (LBW) example with a logistic spatial contextual model.

```
#county level
for (i in 1:m){
yc[i]~dpois(muc[i])
muc[i]<-ec[i]*thc[i]
log(thc[i])<-a0+v[i]+u[i] #convolution model
v[i]~dnorm(0,tauV)
resc[i]<-(yc[i]-muc[i])/sqrt(muc[i])  # standardised residual
}
u[1:m]~car.normal(adjc[],weic[],numc[],tauU)
for( k in 1: SumNumNeigh){weic[k]<-1}
#############################
#PH district level
for (j in 1:p){
yph[j]~dpois(mph[j])
as[j]<-inprod(ec[Label[,j]],thc[Label[,j]])
mph[j]<-as[j]
res[j]<-(yph[j]-mph[j])/sqrt(mph[j])
}
tauV~dgamma(2,0.5)
tauU~dgamma(2,0.5)
}
```

The above code assumes that the public health district (HD) model has a Poisson mean based on a convolution of the expected rate and relative risk from the constituent counties. Note that this fixes the expectation at the HD level but allows the relative risk to be jointly estimated. Nested indexing is used in this example via the binary labeling matrix: Label[,], which assigns the county to one of the 18 health districts. Variants of these joint models are possible where different scale linkage assumptions are made.

Often in surveys, individual participant responses are obtained (via a sampling design which is not primarily spatial) and the respondent is geo-coded to a spatial unit above the resolution of the residential address. The aim may be to use the survey data to estimate the aggregate relative risk of a disease. There is now a growing literature in this area, and two basic approaches have been proposed: (1) use of adjusted outcomes based on sampling design considerations (weighting), and (2) use of unadjusted outcomes and sampling weights used as predictors. The first approach has been proposed by Raghunathan et al. (2007), with modifications in Chen et al. (2014), and Mercer et al. (2014). The second approach was proposed by Vandendijck et al. (2016) (see review in Faes et al., 2016, and Lawson, 2018 Section 7.9). The approach was extended by Watjou et al. (2017). This was extended to multiple disease outcomes by Lawson et al. (2020) in application to metabolic syndrome outcomes in the national health survey within provinces of Chile. The following code (excluding the hyper-prior distributions) displays the model with sampling weight predictor(sw[]) for the binary diabetes outcome:

```
# Model 1 for diabetes
# Including sex, age and sex-age interaction
model{
for(i in 1:m){
y[i]~dbern(p[i])
logit(p[i])<-a0+v[i]+vp[prov[i]]+up[prov[i]]+a[1]*sw[i]+a[2]*age[i]
+a[3]*sex[i]+a[4]*age[i]*sex[i] # vp and up re #provincia level effects
#prov[i] is the index to the provinci
v[i]~dnorm(0,tauv)
pexc[i]<-step(p[i]-0.5)
}
# calculations for diabetes risk ratios
for(k in 1:Np){
pc[k]~dbin(pp[k],n[k])
logit(pp[k])<-ac0+vc[k]+up[k]
vc[k]~dnorm(0,tauvc)
n[k]~dpois(100)
}
# spatial effects
up[1:Np]~car.normal(adj[],wei[],num[],tauup)
for(k in 1:SumNumNeigh){wei[k]<-1}
for (j in 1:Np) #Np is the number of provincias
{vp[j]~dnorm(0,tauvp)}
```

The terms v[i], vp[prov[i]], up[prov[i]]   represent individual and provincia level effects (using nested indexing). The code

```
for(k in 1:Np){
pc[k]~dbin(pp[k],n[k])
logit(pp[k])<-ac0+vc[k]+up[k]
```

```
vc[k]~dnorm(0,tauvc)
n[k]~dpois(100)
}
```

represents the estimation of the provincia level risk proportion (pp[k]). Note that n[k] is only partially observed. Lawson et al. (2020) provide further information about the joint models used to estimate prevalence at provincia level of diabetes, obesity, hypertension, and elevated low density lipoprotein (LDL) from the Chilean national health survey.

18

Infectious Disease Modeling

Most models for the spatial variation of disease are descriptive, in that they simply aim to describe the random variation either by observed predictors or with random effects which can accommodate unobserved confounding. Infectious disease can be treated in like fashion, but the fact that transmission mechanisms are at play in the dynamics of the disease distribution, means that descriptive models cannot hope to allow for accurate prediction as they ignore this effect. Modeling transmission in space and time is a fundamental part of the infectious disease modeling.

Space-time modeling of infectious disease is a fast growing area, in part because of the pandemic of COVID-19 which has spread around the globe in 2020 (Banks et al., 2020).

Here I present a general approach to space-time infection transmission based on the classical susceptible-Infected-Removed (SIR) compartment model (Keeling and Rohani, 2007). The specific model for the case studies was originally proposed by Morton and Finkenstadt (2005). The original model was a SIR time series model applied to weekly measles outbreaks in England and Wales over a period of 20 years.

18.1 Descriptive Modeling: Spatial Flu Modeling

Spatial description of infections is possible. This is essentially a cross-sectional view of a dynamic process. It can be useful, when at a given time, say, a hotspot analysis were to be undertaken. This has been particularly common during the Covid-19 pandemic, when hotspots are identified and further interventions considered. One could consider the temporal marginal distribution of y_{ij} (count of infectives in i th area and j th time). A classic descriptive model would be a convolution model embedded with a data model. Often a Poisson model is assumed as the events are rare. Hence we could assume: $y_{ij} \sim Pois(\mu_{ij})$ and $\mu_{ij} = e_{ij}\theta_{ij}$. Here, e_{ij} could be replaced by the susceptible population (S_{ij}) and the log of θ_{ij} is modeled as $\log(\theta_{ij}) = \beta_0 + v_i + u_i$, where $v_i + u_i$ are a spatial convolution (uncorrelated and correlated heterogeneity, or a Leroux/BYM2 variant). For a given time period we can assess unusual risk areas via posterior exceedance measures. β_0 plays the role of a scaling term (overall rate) and adjust the risk generated by S_{ij}. Hence we could consider $\theta_{ij} = \exp(\beta_0).\lambda_i$ where $log(\lambda_i) = v_i + u_i$ and should be centered on average on zero. Hence $\Pr(\lambda_i > 1)$ could be used to detect unusual increased risk. The posterior average of $\Pr(\lambda_i > 1)$ could be used as a "hot spot" detector. Figure 1.5 displays four county profiles of C+ counts during the flu season. Figure 18.1 displays the exceedance across counties for the SC Flu example for time period 6. There is no time dependence in the model and simply a spatial convolution is fitted. The high exceedance counties of Pickens and Richland, exceed 0.95 and so can be regarded as "significant" hot spots. Charleston county is elevated but does not reach the 0.95 category. It is also notable that these counties show up as positive outliers in the UH effect also. The correlated effect suggests areas of higher risk in the north west mainly.

18.2   Mechanistic Modeling: SC Flu Season 2004/2005 and Covid-19 (2020)

Transmission dynamics are fundamental to the mechanisms of infection. In fact most infectious disease models formed as differential equations are purely temporally dynamic (see e.g. Keeling and Rohani, 2007). However it is clear that spatial evidence for infection spread and differentials between different places WRT new infections could be very important not least from a public Health surveillance perspective. In the following, two examples of spatio-temporal dependence modeling are provided. The first is a single season of influenza with counts at the county level in South Carolina. The data are biweekly counts of influenza C+ notifications during the 2004–2005 flu season. The second consists of daily data on counts of Covid-19 positive tests during the period of April 2nd until June 29th 2020 (88 days) also in the counties of South Carolina.

18.2.1   SC Flu Season 2004–2005 Modeling

Essentially, the SC flu example consists of a matrix of counts 46 rows (counties) and 13 columns (time periods). An analysis of these data is reported in Lawson (2018), Chapter 14. The analysis reported was carried out using OpenBUGS. The code reported here is for a nimble implementation. The outcome is cpos[i,j], and the initial susceptible population is susint[i]. The model implemented is essentially a spatial extension of the Morton and Finkenstadt (2005) (M&F) time series susceptible-infected-removed (SIR) model. In this model the outcome is assumed to have a Poisson distribution with dependence on previous levels of the count. Hence, y_{ij} is the C+ve count and it is assumed

FIGURE 18.1

Single time period (time 6) posterior model estimates of UH effect (left), the ICAR effect (middle) and exceedance probability (right)

that

$$y_{ij} \sim Pois(\mu_{ij})$$
$$\mu_{ij} = S_{ij} f(y_{i,j-1} \dots)$$

where the mean level is a function of the susceptible population S_{ij} and a function of the previous count within that area $y_{i,j-1}$. The function $f(y_{i,j-1} \dots)$ could have a variety of forms and additional effects could be included in the formulation. For example, with the usual log transformation we could have $\log(\mu_{ij}) = \log(S_{ij}) + \log f(y_{i,j-1} \dots)$ and $\log f(y_{i,j-1} \dots) = \log(y_{i,j-1}) + b_0 + b_{1i}$ where b_0 is a transmission rate and b_{1i} could be a random effect (possibly an ICAR effect). In the code below bet0 is the log-scale transmission rate and b1[] has an ICAR prior distribution. This spatial effect is included to allow for spatially structured confounding. An accounting equation is also included which updates the susceptible pool over time: $S_{ij} = S_{i,j-1} - y_{i,j-1} - R_{i,j-1}$. Removal (R_{ij}) is assumed to have a fixed rate and is a function of current infectives.

```
Fullmodel<-nimbleCode({
for (i in 1:M){
rem[i,1]<-0
susc[i,1]<-susint[i]
muc[i,1]<-0.00001*susc[i,1]
cpos[i,1]~dpois(muc[i,1])
}
for (i in 1:M){
for (j in 2: T){
rem[i,j]<-betaR*cpos[i,j]
susc[i,j]<-susc[i,j-1]-cpos[i,j-1]-rem[i,j-1]
cpos[i,j]~dpois(muc[i,j])
log(muc[i,j])<-bet0+log(susc[i,j]+0.001)+log(cpos[i,j-1]+0.001)+b1[i]
}
muct1[i]<-muc[i,1]
muct2[i]<-muc[i,2]
muct3[i]<-muc[i,3]
muct4[i]<-muc[i,4]
muct5[i]<-muc[i,5]
muct6[i]<-muc[i,6]
muct7[i]<-muc[i,7]
muct8[i]<-muc[i,8]
muct9[i]<-muc[i,9]
muct10[i]<-muc[i,10]
muct11[i]<-muc[i,11]
muct12[i]<-muc[i,12]
```

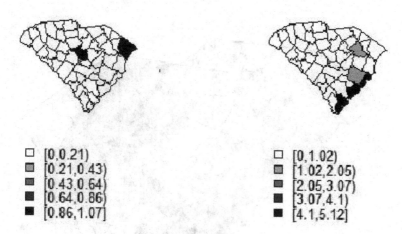

FIGURE 18.2
Posterior mean estimates of risk at time period 6 (left) and 12 (right) during
the 13 period flu season 2004-2005 in South Carolina.

```
muct13[i]<-muc[i,13]
}
for (j in 1:T){
mucrich[j]<-muc[40,j]
mucchar[j]<-muc[10,j]
muchor[j]<-muc[26,j]
mucbea[j]<-muc[7,j]}
b1[1:46] ~dcar_normal(adj[1:L], weights[1:L], num[1:M], tau.b1,zero_mean=1)
for(k in 1:L) {weights[k] <- 1 }
bet0~dflat()
tau.b1~dgamma(0.01,0.01)
betaR<-0.001})
```

In this code the profiles of four selected counties (Richland, Charleston,
Horry, and Beaufort) are set up and fixed time mean estimates are stored in
muct1[] - muct13[]. For a converged sample of 20000 iterations (with burnin
10000) the resulting sample yielded the following maps and profiles. Figure
18.2 displays the posterior estimates of risk at time period 6 and time
period 12 during the flu season. Figure 18.3 displays the residual posterior
mean ICAR effect for the flu example. It is notable that there is some positive
clustering in the rural areas in the south west and north east of the state and
in these areas the model does not fit so well. It is notable that the urban
counties are much better modeled in general.

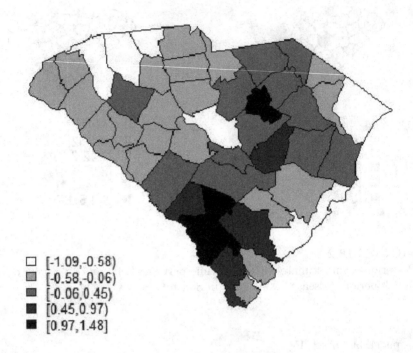

[-1.09,-0.58)
[-0.58,-0.06)
[-0.06,0.45)
[0.45,0.97)
[0.97,1.48]

FIGURE 18.3
Posterior mean ICAR effect for the SC flu season example.

Finally, Table 18.1 displays the four county posterior mean risk profiles over the 13 period season. The displays includes the 95% credible interval for the risk (shaded areas). It is notable that different counties in different areas experience the epidemic peaks at different times with the coastal counties experiencing later season up turns in risk compared to the urban areas of Charleston and Richland.

18.2.2   SC county-level Covid-19 Modeling

During the pandemic of Covid-19 in 2020 various attempts have been made to model the development of the virus progression. Early predictive modeling focused on the computation of R_0 and on making ball-park predictions of the temporal dimension of the outbreak. Few attempts have been made to consider modeling the spatial spread of the virus, although various attempts have been made to test for clustering or "hot spots"(see e.g. Gomes et al., 2020, Desjardins et al., 2020). An exception is Zhou et al. (2020) who employ susceptible infected removed (SIR) models based on differential equations. They

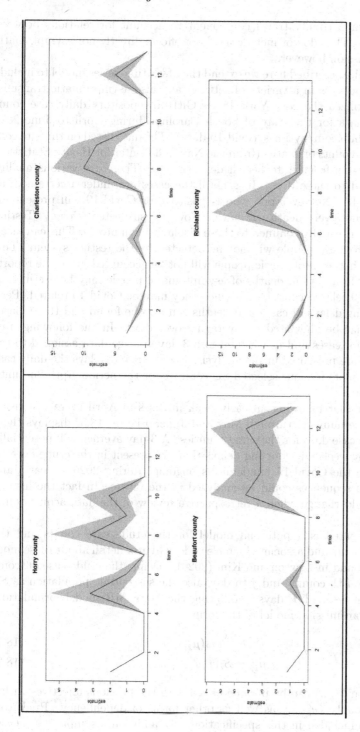

Table 18.1
Posterior mean profiles for 4 selected counties for the SC flu example.

include an inter-county travel component to account for relations between spatial units. They do not include neighbor effects directly nor asymptomatic case transmission however.

In the model described here we extend the M&F time series model to include spatial components in a variety of settings and also asymptomatic transmission. We utilized the New York Times GitHub repository daily case count and death data for the state of South Carolina during April to June 2020 (https://github.com/nytimes/covid-19-data). The data listed on this site consists of daily counts of deaths (from the National Center for Health Statistics) and case counts from state health departments. There are various quality issues related to these data. In general the cases are under-ascertained, as are the deaths. Not everyone who is positive for Covid-19 will present for testing and not every death will have a Covid-19 suspected etiology. Testing for Covid-19 is usually confined to those displaying symptoms. This can mean that asymptomatic people will not be detected by the testing system. This also means that asymptomatic people will not be accounted for in the reporting system. In addition, deaths of asymptomatic people may be ascribed to non-Covid-19 related causes, even when they may be Covid-19 related. Periodic reporting delays of cases and deaths can also be found and these biases can somewhat be alleviated by averaging over days. In the following, raw count data for cases and deaths and also 3-day average data assigned to the central day, are presented in the analysis. Figure 18.4 displays the daily case and death counts for four selected SC counties for the period April 2nd until June 29th 2020.

It is clear that following an early peak in cases in April there is a rising trend on case numbers through May and June. Figure 18.5 displays the 3 day average case data for the same counties. A 3-day average will potentially allow for any weekend reporting delays that are present in the crude data.

Of course the Covid-19 pandemic is ongoing (during 2020 at least) and further data sequences could be analyzed in the future. In fact the historical data could change when a retrospective review of the data acquisition is attempted.

Here presented is a potential model for the crude case counts and the smoothed counts and associated nimble code. Greater detail about these models can be found in Lawson and Kim (2020). Define the crude case outcome as y_{ij}^c in the i th county and j th day, also the susceptible population as S_{ij} with m counties and T days. Following the M&F SIR model formulation, assume a transmission model of the form

$$y_{ij}^c \sim Pois(\mu_{ij}) \tag{18.1}$$

$$\mu_{ij} = S_{ij} f(y_{i,j-1}^c \cdots \cdots) \tag{18.2}$$

The function $f(\cdots.)$ can be parameterized with the previous infectives within the same county $(y_{i,j-1}^c)$ as well as other forms of dependency. Predictors can also be included in this specification. In addition, asymptomatic cases

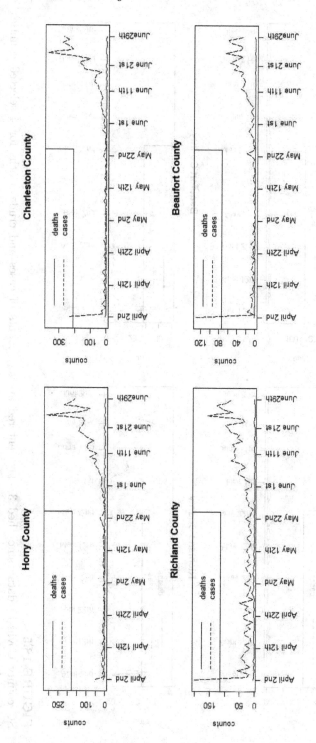

FIGURE 18.4
SC Covid-19 example: daily cases and deaths April 2nd to June 29th: 4 selected counties.

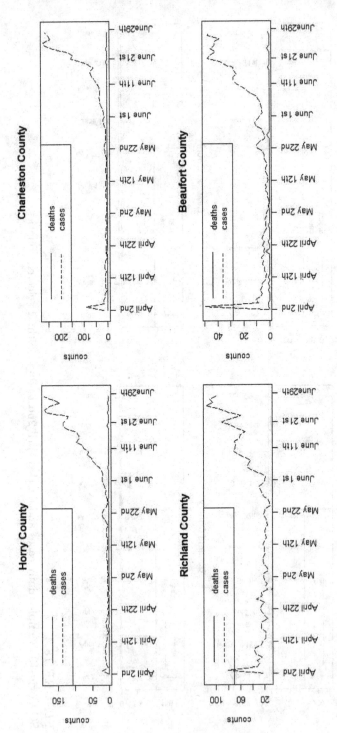

FIGURE 18.5

SC county Covid-19 data: smoothed 3-day centered average counts of cases and crude deaths for 4 selected counties (publication pending).

(y_{ij}^{as}) can be added to the transmission potential. Asymptomatic cases can be regarded as a latent process, which is largely unobserved. As such it is relevant to model these as they affect the infective load, but it is computationally very expensive to estimate this as a latent ST process. Instead a simple dependence is assumed whereby $y_{ij}^{as} = \beta_{as} y_{ij}^{c}$. Current estimates of the asymptomatic rate range from 0.25 up to 0.5. In Lawson and Kim (2020) a different rate is assumed in different model runs to assess goodness of fit.

In addition to the transmission model an accounting equation must be updated:

$$S_{ij} = S_{i,j-1} - I_{i,j-1}^{c} - R_{i,j-1},$$

where $I_{i,j-1}^{c} = y_{i,j-1}^{c} + y_{i,j-1}^{as}$. Removal (R_{ij}) is assumed to have two components: recovery as a fixed rate and is a function of current infectives (as this is not directly available), and death removal as a death count.

Note that the crude case rate (y_{ij}^{c}) is dependent ultimately on the number of tests undertaken. An improved version might assume a binomial transmission model with $y_{ij}^{c} \sim bin(p_{ij}, n_{ij})$ where n_{ij} is the number of tests performed. Lagged effects could also be included of course. The modeling could focus on the probability of a positive test p_{ij}. However, currently the number of daily tests is not publicly available and so we must fall back on the Poisson assumption.

18.2.3 Model Variants

Coupled with the base model 18.1, different models can be specified for $f(y_{i,j-1}^{c}.....)$. A basic form is

$$log(f(..)) = \beta_0 + \beta_1 \log(I_{i,j-1}^{c}) + b1_i, \tag{18.3}$$

where $b1_i$ is a spatially referenced random effect term to allow for confounding. An ICAR prior distribution model is assumed for this term.

Further models with different dependencies have been examined:

$$log(f(..)) = \beta_0 + \beta_1 \log(I_{i,j-1}^{c} + \sum_{l \in \delta_i} I_{l,j-1}^{c}) + b1_i \tag{18.4}$$

where $\sum_{l \in \delta_i} I_{l,j-1}^{c}$ is the sum of neighboring (adjacent) region infectives.

$$log(f(..)) = \beta_0 + \beta_1 \log(I_{i,j-1}^{c}) + \beta_2 x_i + b1_i \tag{18.5}$$

where x_i is a county level measure of deprivation. In this case % population under the poverty line.

$$log(f(..)) = \beta_{0j} + \beta_1 \log(I_{i,j-1}^{c}) + \beta_2 x_i + b1_i \tag{18.6}$$

where β_{0j} is now time dependent.

$$log(f(..)) = \beta_{0i} + \beta_1 \log(I_{i,j-1}^{c}) + \beta_2 x_i + v_i \tag{18.7}$$

where β_{0i} the overall transmission rate is spatially dependent and v_i is an uncorrelated random effect. Many other variants of these models could be examined and in particular predictors of various kinds could be examined. Using DIC as a goodness of fit criterion, for the crude daily data, model 3 with % poverty covariate and 0.25 asymptomatic rate fit the data best. In fact, the inclusion of the poverty covariate and spatial confounding were found to lead to the lowest DIC models. Figure 18.6 displays the model 3 posterior mean estimates for Greenville and Charleston counties. Figure 18.7 displays the one step prediction for June 30th from this model. It is notable that the predictions for Charlston, Richland, and Horry and Greenville have large upper 95% limits.

18.2.3.1   Smoothed Data

Using the 3 - day smoothed data, yielded a slightly different picture. The centered smoothed average was assumed to be log normally distributed (as it is continuous):

$$y_{ij}^{sm} = \exp(y_{ij}^{*})$$
$$y_{ij}^{*} \sim N(\mu_{ij}, \tau_y^{-1}).$$

A range of models for μ_{ij} were assumed, as before. In this case, the best model based on DIC was the model 3 with the % population under poverty variable, constant transmission intercept (β_0) and ICAR random effect. Figure 18.8 displays the one step prediction for June 30th based on the smoothed data. It is notable that the Beaufort and Lexington counties have a reduced prediction range compared to the prediction for the crude daily data (Figure 18.7), whereas Richland has a larger range.

As part of the output from these rich Bayesian models, it is also possible to examine maps of posterior risk estimates and to compute exceedance maps. To that end, Figure 18.9 displays a sequence of six maps of daily posterior mean risk during an earlier pandemic period (day 10, 40, 46, 50, 60, 70: January 22nd to April 12th). Notice that the maps look similar except that the ranges of risk increase considerably over time. For the first 3 dates the risk is negligible but then the later dates have much increased risk.

Finally it is also possible to consider evidence for hot spots during a pandemic of this kind. The usual exceedance probability computation can be used. As an example, I have calculated the exceedance for two dates during the later April 2nd to June 29th period: day 45 (May 16th) and day 88 (June 29th). I used the overall average risk across all regions and times as threshold. Absolute thresholds could also be assumed of course. Figure 18.10 displays the day 45 exceedance, while Figure 18.11 displays the day 88 (June 29th) result. It is clear that the day 45 hotspots of significantly elevated risk (counties with exeeedance above 0.95) have appeared in the upstate and central rural counties as well as Richland and Charleston. By Day 88, however

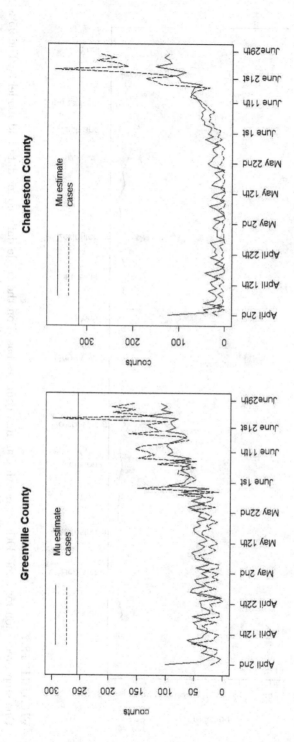

FIGURE 18.6
Charleston and Greenville counties: posterior mean risk estimates (μ_{ij}) for model 3 for crude daily data (publication pending).

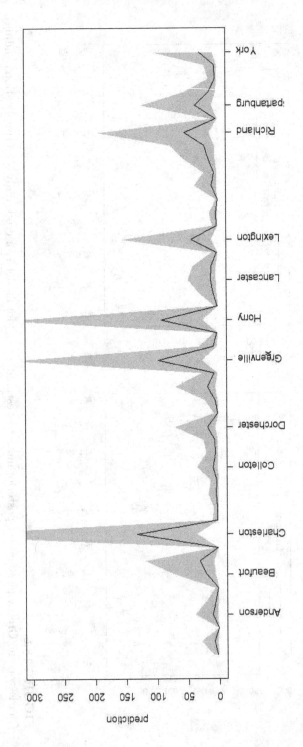

FIGURE 18.7

One step prediction for June 30th for South Carolina counties based on the crude daily count data and model 3. The 95% credible intervals are denoted by the shaded areas (publication pending).

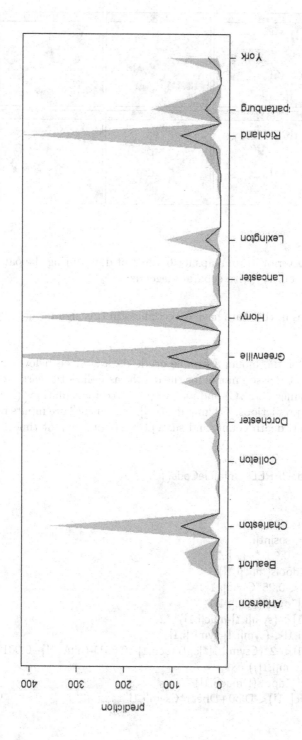

FIGURE 18.8

3-day smoothed data: one step prediction with 95% credible intervals for SC counties (publication pending).

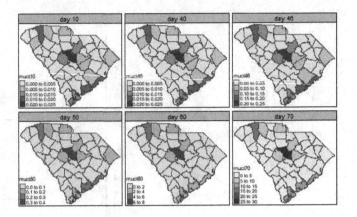

FIGURE 18.9
SC county posterior mean maps for 6 different days during the pandemic (day 10, 40,46, 50, 60, 70) with separate legends.

many counties in the state have exceedances in this significantly elevated risk category.

The code for the model 3 for the crude data is given below. This version includes a joint Poisson model for the deaths as well as the case model. It also includes deviance (LdevC) and mean square error calculations (MSE, MSPE) and one step predictions. Csym[], deaths[], and susint[] are inputs representing the case count, death count, and susceptible population at time 1.

```
COVModel3PRED<-nimbleCode({
for (i in 1:M){
remc[i,1]<-0
remD[i,1]<-0
susc[i,1]<-susint[i]
muc[i,1]<-0.001*susc[i,1]
sym[i,1]~dpois(muc[i,1])
asym[i,1]<-0.25*Csym[i,1]
symP[i,1]~dpois(muc[i,1])
SEloss[i,1]<-(sym[i,1]-muc[i,1])**2
SPEloss[i,1]<-(sym[i,1]-symP[i,1])**2
LdevC[i,1]<-–2*(Csym[i,1]*log(muc[i,1]+0.001)-(muc[i,1]+0.001)-
lfactorial(Csym[i,1]))
deaths[i,1]~dpois(Dmuc[i,1])
log(Dmuc[i,1])<-Dlp0+Dbet1*Csym[i,1]
```

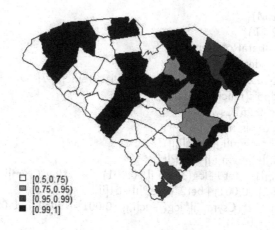

FIGURE 18.10
Posterior exceedance probability for day 45 (May 16th) with overall mean risk threshold.

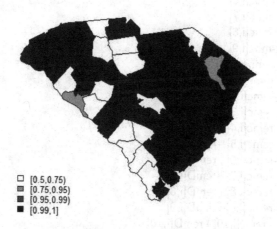

FIGURE 18.11
Posterior average exceedance for day 88 (June 29th) using the overall risk mean as threshold.

```
      }
      for (i in 1:M){
      for (j in 2: T){
      remc[i,j]<-betaRc*Csym[i,j]
      remD[i,j]<-deaths[i,j]
      susc[i,j]<-susc[i,j-1]-Csym[i,j-1]-asym[i,j-1]-remc[i,j-1]-remD[i,j-1]
      sym[i,j]~dpois(muc[i,j])
      asym[i,j]<-0.25*Csym[i,j]
      symP[i,j]~dpois(muc[i,j])
      SEloss[i,j]<-(sym[i,j]-muc[i,j])**2
      SPEloss[i,j]<-(sym[i,j]-symP[i,j])**2
      log(muc[i,j])<-bet0+log(susc[i,j]+0.001)+bet1*log(Csym[i,j-1]+0.001)+
log(asym[i,j-1]+0.001)+bet2*percP[i]+b1[i]
      LdevC[i,j]<-2*(Csym[i,j]*log(muc[i,j]+0.001)-(muc[i,j]+0.001)-
lfactorial(Csym[i,j]))
      deaths[i,j]~dpois(Dmuc[i,j])
      log(Dmuc[i,j])<-Dlp0+Dbet1*Csym[i,j]+Dbet2*TSym[i,j-1]
      }
      muct1[i]<-muc[i,1]
      muct2[i]<-muc[i,2]
      muct3[i]<-muc[i,3]
      muct4[i]<-muc[i,4]
      muct5[i]<-muc[i,5]
      muct6[i]<-muc[i,6]
      muct7[i]<-muc[i,7]
      muct8[i]<-muc[i,8]
      muct9[i]<-muc[i,9]
      muct10[i]<-muc[i,10]
      remt1[i]<-remc[i,1]+remD[i,1]
      remt2[i]<-remc[i,2]+remD[i,2]
      remt3[i]<-remc[i,3]+remD[i,3]
      remt4[i]<-remc[i,4]+remD[i,4]
      remt5[i]<-remc[i,5]+remD[i,5]
      remt6[i]<-remc[i,6]+remD[i,6]
      remt7[i]<-remc[i,7]+remD[i,7]
      remt8[i]<-remc[i,8]+remD[i,8]
      remt9[i]<-remc[i,9]+remD[i,9]
      remt10[i]<-remc[i,10]+remD[i,10]
      sust1[i]<-susc[i,1]
      sust2[i]<-susc[i,2]
      sust3[i]<-susc[i,3]
      sust4[i]<-susc[i,4]
      sust5[i]<-susc[i,5]
```

```
sust6[i]<-susc[i,6]
sust7[i]<-susc[i,7]
sust8[i]<-susc[i,8]
sust9[i]<-susc[i,9]
sust10[i]<-susc[i,10]
}
for (j in 1:T){
mucrich[j]<-muc[40,j]
mucchar[j]<-muc[10,j]
muchor[j]<-muc[26,j]
mucbea[j]<-muc[7,j]
mucker[j]<-muc[28,j]
mucand[j]<-muc[4,j]
mucsum[j]<-muc[43,j]
mucflor[j]<-muc[21,j]
mucgreen[j]<-muc[23,j]
}
b1[1:46] ~dcar_normal(adj[1:L], wei[1:L], num[1:M], tau.b1,zero_mean=1)
 for(k in 1:L) {wei[k] <- 1 }
for ( j in 1:T){
MSE[j]<-mean(SEloss[1:M,j])
MSPE[j]<-mean(SPEloss[1:M,j])
}
DevC<-sum(LdevC[1:M,1:T])
############ one step forecast ###############
for (k in 1:M) {
remc1[k] <- betaRc * Csym[k,T]
remD1[k] <- deaths[k,T]
susc1[k] <- susc[k,T] - Csym[k,T]- (0.25*Csym[k,T]) - remc1[k] - remD1[k]
log(muc1[k]) <- bet0 + log(susc1[k] + .001) + bet1*log(Csym[k,T] +
.001) +
        log(0.25 * Csym[k,T] + 0.001) +bet2*percP[k]+ b1[k]
m1[k]<-log(muc1[k]+0.001)-log(susc1[k]+0.001)
Nmu[k]~dnorm(m1[k],sigN)
yP1[k] <-ceiling(exp(log(susc1[k]+0.001)+Nmu[k]))}
######## yP1 is prediction for next day: June 30th #########
Dlp0~dnorm(0,tauD0)
Dbet1~dnorm(0,tauD1)
Dbet2~dnorm(0,tauD2)
tauD0~dgamma(2,0.5)
tauD1~dgamma(2,0.5)
tauD2~dgamma(2,0.5)
sigN~dgamma(2,0.5)
```

```
bet0~dnorm(0,tau0)
tau0~dgamma(2,0.5)
bet1~dnorm(0,tau1)
tau1~dgamma(2,0.5)
tau.b1~dgamma(0.01,0.01)
betaRc<-0.1
bet2~dnorm(0,tau2)
tau2~dgamma(2,0.5)
})
```

References

Abellan, J., S. Richardson, and N. Best (2008). Use of space-time models to investigate the stability of patterns of disease. *Environmental Health Perspectives 116*, 1111–1118. doi:10.1289/ehp.10814.

Aregay, M., A. B. Lawson, C. Faes, and R. Kirby (2014). Bayesian multiscale modeling for aggregated disease mapping data. In *Proceedings of the Third ACM SIGSPATIAL International Workshop on the Use of GIS in Public Health*, HealthGIS '14, New York, NY, USA, pp. 45–48. ACM.

Aregay, M., A. B. Lawson, C. Faes, R. Kirby, R. Carroll, and K. Watjou (2016a). Multiscale spatiotemporal models for aggregated small area health outcomes. *Environmetrics*. to appear.

Aregay, M., A. B. Lawson, C. Faes, and R. S. Kirby (2015). Bayesian multiscale modeling for aggregated disease mapping data. *Statistical Methods in Medical Research 26(6)*, 2726–2742.

Aregay, M., A. B. Lawson, C. Faes, R. S. Kirby, R. Carroll, and K. Watjou (2016b). Spatial mixture multiscale modeling for aggregated health data. *Biometrical Journal 58*(5), 1091–1112.

Baddeley, A. and R. Turner (2000). Practical maximum pseudolikelihood for spatial point patterns. *Australia and New Zealand Journal of Statistics 42*, 283–322.

Baer, D. R. and A. B. Lawson (2019). Evaluation of Bayesian multiple stage estimation under spatial car model variants. *Journal of Statistical Computation and Simulation 89*(1), 98–144.

Baer, D. R., A. B. Lawson, and J. Joseph (2020). Joint spacetime Bayesian disease mapping via quantification of disease risk association. *Statistical Methods in Medical Research*. to appear.

Balakrishnan, S. and D. Madigan (2006). Algorithms for sparse linear classifiers in the massive data setting. *Journal of Machine Learning Research 1*.

Banerjee, S. (2016). Spatial survival models. In A. B. Lawson, S. Banerjee, R. Haining, and L. Ugarte (Eds.), *Handbook of Spatial Epidemiology*. New York: CRC Press.

Banerjee, S., B. Carlin, and A. Gelfand (2014). *Hierarchical Modelling and Analysis for Spatial Data* (2 ed.). New York: CRC Press.

Banerjee, S., B. P. Carlin, and A. E. Gelfand (2004). *Hierarchical Modeling and Analysis for Spatial Data.* London: Chapman and Hall/CRC Press.

Banerjee, S., A. Gelfand, A. Finley, and H. Sang (2008). Gaussian predictive process models for large spatial data sets. *Journal of the Royal Statistical Society 70*, 825–848.

Banerjee, S., M. M. Wall, and B. P. Carlin (2003). Frailty modeling for spatially correlated survival data, with application to infant mortality in Minnesota. *Biostatistics 4*, 123–142.

Banks, D., S. Ellenberg, T. Fleming, M. E. Halloran, A. Lawson, and L. Waller (2020, 5). A conversation about covid-19 with biostatisticians and epidemiologists. *Harvard Data Science Review.* https://hdsr.mitpress.mit.edu/pub/9qtvk9qi.

Barbieri, M. and J. Berger (2004). Optimal predictive model selection. *Annals of Statistics 32*, 870–897.

Bastos, L. and D. Gamerman (2006). Dynamical survival models with spatial frailty. *Lifetime Data Analysis 12*, 441–460.

Berchuck, S., M. Janko, F. Medeiros, W. Pan, and S. Mukherjee (2019). Bayesian non-parametric factor analysis for longitudinal spatial surfaces. *arxiv:1911.04337v1.* prepub.

Berman, M. and T. R. Turner (1992). Approximating point process likelihoods with GLIM. *Applied Statistics 41*, 31–38.

Bernardinelli, L., D. G. Clayton, C. Pascutto, C. Montomoli, M. Ghislandi, and M. Songini (1995). Bayesian analysis of space-time variation in disease risk. *Statistics in Medicine 14*, 2433–2443.

Bernardo, J. M. and A. F. M. Smith (1994). *Bayesian Theory.* New York: Wiley.

Besag, J. and P. J. Green (1993). Spatial statistics and Bayesian computation. *Journal of the Royal Statistical Society, Series B 55*, 25–37.

Besag, J. and J. Tantrum (2003). Likelihood analysis of binary data in space and time. In P. Green, N. Hjort, and S. Richardson (Eds.), *Highly Structured Stochastic Systems*, Chapter 9A, pp. 289–295. Oxford University Press.

Besag, J., J. York, and A. Mollié (1991). Bayesian image restoration with two applications in spatial statistics. *Annals of the Institute of Statistical Mathematics 43*, 1–59.

Best, N., S. Richardson, and A. Thomson (2005). A comparison of Bayesian spatial models for disease mapping. *Statistical Methods in Medical Research 14*, 35–59.

Blangiardo, M., M. Cameletti, G. Baio, and H. Rue (2013). Spatial and spatio-temporal models with R-INLA. *Spatial and Spatio-temporal Epidemiology 4*, 33–49.

Botella-Rocamora, P., M. Martinez-Beneito, and S. Banerjee (2015). A unifying modeling framework for highly multivariate disease mapping. *Statistics in Medicine 34*(9), 1548–1559.

Breslow, N. and D. G. Clayton (1993). Approximate inference in generalised linear mixed models. *Journal of the American Statistical Association 88*, 9–25.

Brooks, S., A. Gelman, G. Jones, and X. Meng (Eds.) (2011). *Handbook of Markov Chain Monte Carlo*. New York: CRC Press.

Brooks, S. and A. E. Gelman (1998). General methods for monitoring convergence of iterative simulations. *Journal of Computational and Graphical Statistics 7*, 434–455.

Brooks, S. P. (1998). Quantitative convergence assessment for MCMC via Cusums. *Statistics and Computing 8*, 267–274.

Calder, C. A. (2007). Dynamic factor process convolution models for multivariate space–time data with application to air quality assessment. *Environmental and Ecological Statistics 14*(3), 229–247.

Calder, C. A. (2008). A dynamic process convolution approach to modeling ambient particulate matter concentrations. *Environmetrics 19*(1), 39–48.

Calder, C. A., C. Holloman, and D. Higdon (2002). Exploring space-time structure in ozone concentration using a dynamic process convolution model. In *Bayesian Case Studies VI*. New York: Springer.

Carlin, B. and S. Chib (1995). Bayesian model choice via Markov Chain Monte Carlo methods. *Journal of the Royal Statistical Society. Series B (Methodological) 57*, 473–484.

Carlin, B. P. and S. Banerjee (2003). Hierarchical multivariate CAR models for spatio-temporally correlated survival data. In J. B. Bernardo, D. A. P., J. O. Berger, and M. West (Eds.), *Bayesian Statistics 7*. Oxford: Oxford University Press.

Carlin, B. P. and T. Louis (2000). *Bayes and Empirical Bayes Methods for Data Analysis* (2nd ed.). Chapman and Hall/CRC press.

Carroll, R., A. B. Lawson, C. Faes, R. Kirby, and M. Aregay (2015). Comparing INLA and OpenBUGS for hierarchical Poisson modeling in disease mapping. *Spatial and Spatio-temporal Epidemiology 14–15*, 45–54.

Carroll, R., A. B. Lawson, C. Faes, R. Kirby, M. Aregay, and K. Watjou (2016). Bayesian model selection methods in modeling small area colon cancer incidence. *Annals of Epidemiology 26*, 43–49.

Carroll, R., A. B. Lawson, C. Faes, R. S. Kirby, M. Aregay, and K. Watjou (2017). Space-time variation of respiratory cancers in South Carolina: A flexible multivariate mixture modeling approach to risk estimation. *Ann Epidemiol*. special issue.

Carroll, R., D. Ruppert, L. Stefanski, and C. Crainiceanu (2006). *Measurement Error in Nonlinear Models* (2 ed.). New York: Chapman & Hall.

Cassella, G. and E. I. George (1992). Explaining the Gibbs Sampler. *The American Statistician 46*, 167–174.

Celeux, G., F. Forbes, C. Robert, and M. Titterington (2006). Deviance information criteria for missing data models. *Bayesian Analysis 1*, 651–674.

Chan, T. C., C. C. King, M. Y. Yen, P. Chiang, C. S. Huang, and C. K. Hsiao (2010). Probabilistic daily ILI syndromic surveillance with a spatio-temporal Bayesian hierarchical model. *PLoS ONE 5(7)*, e11626.

Chang, H. (2016). Data assimilation for environmental pollution fields. In A. B. Lawson, S. Banerjee, R. Haining, and L. Ugart (Eds.), *Handbook of Spatial Epidemiology*, pp. 289–302. New York: CRC Press.

Chen, C., J. Wakefield, and T. Lumley (2014). The use of sample weights in Bayesian hierarchical models for small area estimation. *Spatial and Spatio-temporal Epidemiology 11*, 33–43.

Chen, M., Q. Shao, and J. Ibrahim (2000). *Monte Carlo Methods in Bayesian Computation*. New York: Springer Verlag.

Choi, J. and A. B. Lawson (2011). Evaluation of Bayesian spatial-temporal latent models in small area health data. *Environmetrics 22(8)*, 1008–1022.

Clayton, D. G. and J. Kaldor (1987). Empirical Bayes estimates of age-standardised relative risks for use in disease mapping. *Biometrics 43*, 671–691.

Congdon, P. (2010). *Applied Bayesian Hierarchical Methods*. New York: CRC Press.

Cooner, F., S. Banerjee, and M. McBean (2006). Modelling geographically referenced survival data with a cure fraction. *Statistical Methods in Medical research 15(4)*, 307–324.

Corberán-Vallet, A. (2012). Prospective surveillance of multivariate spatial disease data. *Statistical Methods in medical Research 21*(5), 457–477. DOI:10.1177/0962280212446319.

Cressie, N. A. C. (1993). *Statistics for Spatial Data* (revised ed.). New York: Wiley.

Cressie, N. A. C. and N. H. Chan (1989). Spatial modelling of regional variables. *Journal of the American Statistical Association 84*, 393–401.

Cressie, N. A. C. and C. Wikle (2011). *Statistics for Spatio-temporal Data*. New York: Wiley.

Dabney, A. and J. Wakefield (2005). Issues in the mapping of two diseases. *Statistical Methods in Medical Research 14*, 83–112.

Dellaportas, P., J. Forster, and I. Ntzoufras (2002). On Bayesian model and variable selection using MCMC. *Statistics and Computing 12*, 27–36.

Desjardins, M., A. Hohl, and E. Delmelle (2020). Rapid surveillance of covid-19 in the united states using a prospective space-time scan statistic: Detecting and evaluating emerging clusters. *Applied Geography 118*, 102202.

Dey, D., M.-H. Chen, and H. Chang (1997). Bayesian approach for nonlinear random effects models. *Biometrics 53*, 1239–1252.

Diggle, P. and B. Rowlingson (1994). A conditional approach to point process modelling of elevated risk. *Journal of the Royal Statistical Society A 157*, 433–440.

Diggle, P. J. (1990). A point process modelling approach to raised incidence of a rare phenomenon in the vicinity of a prespecified point. *Journal of the Royal Statistical Society 153*, 349–362.

Diggle, P. J., J. Tawn, and R. Moyeed (1998). Model-based Geostatistics. *Journal of the Royal Statistical Society C 47*, 299–350.

Diva, U., D. K. Dey, and S. Banerjee (2008). Parametric models for spatially correlated survival data for individuals with multiple cancers. *Statistics in Medicine 27*(12), 2127–2144.

Dominici, F., L. Sheppard, and M. Clyde (2003). Health effects of air pollution. *International Statistical Review 71*, 243–276.

Eberley, L. and B. P. Carlin (2000). Identifiability and convergence issues for Markov Chain Monte Carlo fitting of spatial models. *Statistics in Medicine 19*, 2279–2294.

Elliott, P., G. Shaddick, J. Wakefield, C. Hoogh, and D. Briggs (2007). Long-term associations of outdoor air pollution with mortality in Great Britain. *Thorax 62*, 1088–1094.

Faes, C., Y. Vandendijk, and A. B. Lawson (2016). Spatial health survey data. In A. B. Lawson, S. Banerjee, R. Haining, and L. Ugart (Eds.), *Handbook of Spatial Epidemiology*, pp. 563–574. New York: CRC Press.

Fernandez, C. and P. Green (2002). Modelling spatially correlated data via mixtures: A Bayesian approach. *Journal of the Royal Statistical Society B 64*, 805–826.

Ferreira, J., D. Denison, and C. Holmes (2002). Partition modelling. In A. B. Lawson and D. Denison (Eds.), *Spatial Cluster Modelling*, Chapter 7, pp. 125–145. New York: CRC press.

Fisher, G. and A. B. Lawson (2020). Bayesian modeling of georeferenced cancer survival. *Annals of Cancer Epidemiology 4*(0).

Fonseca, T. C. O. and M. A. R. Ferreira (2017). Dynamic multiscale spatiotemporal models for Poisson data. *Journal of the American Statistical Association 112*(517), 215–234.

Gamerman, D. and H. Lopes (2006). *Markov Chain Monte Carlo: Stochastic Simulation for Bayesian Inference*. New York: CRC Press.

Gelfand, A. and D. Dey (1994). Bayesian model choice: Asymptotics and exact calculations. *Journal of the Royal Statistical Society B 56*, 501–514.

Gelfand, A., P. Diggle, M. Fuentes, and P. Guttorp (Eds.) (2010). *Handbook of Spatial Statistics*. New York: CRC Press.

Gelfand, A. and S. Ghosh (1998). Model choice: A minimum posterior predictive loss approach. *Biometrika 85*, 1–11.

Gelfand, A. and P. Vounatsou (2003). Proper multivariate conditional autoregressive models for spatial data. *Biostatistics 4*, 11–25.

Gelman, A. (2006). Prior distributions for variance parameters in hierarchical models. *Bayesian Analysis 1*, 515–533.

Gelman, A., J. B. Carlin, H. S. Stern, and D. Rubin (2004). *Bayesian Data Analysis*. London: Chapman and HAll/CRC press.

Gelman, A., J. Huang, and A. Vehtari (2014). Understanding predictive information criteria in Bayesian models. *Statistics and Computing 24*, 997–1016.

Gelman, A. and K. Shirley (2011). Inference and monitoring convergence. In S. Brooks, A. Gelman, G. Jones, and X. Meng (Eds.), *Handbook of Markov Chain Monte Carlo*, Chapter 6. New York: CRC Press.

Gelman, A. E. and D. Rubin (1992). Inference from iterative simulation using multiple sequences (with discussion). *Statistical Science 7*, 457–511.

Geweke, J. (1992). Evaluating the accuracy of sampling-based approaches to the calculation of posterior moments. In J. Bernardo, J. Berger,

A. Dawid, and A. Smith (Eds.), *Bayesian Statistics 4*. Oxford University Press.

Ghosh, M. and J. N. K. Rao (1994). Small area estimation: An appraisal. *Statistical Science 9*, 55–93.

Gilks, W. R., D. G. Clayton, D. J. Spiegelhalter, N. G. Best, A. J. McNeil, L. D. Sharples, and A. J. Kirby (1993). Modelling complexity: Applications of Gibbs sampling in medicine. *Journal of the Royal Statistical Society B 55*, 39–52.

Gilks, W. R., S. Richardson, and D. J. Spiegelhalter (Eds.) (1996). *Markov Chain Monte Carlo in Practice*. London: Chapman and Hall.

Goldstein, H. and A. Leyland (2001). Further topics in multilevel modelling: Context and composition. In A. Leyland and H. Goldstein (Eds.), *Multilevel Modelling of Health Statistics*, Chapter 12.4, pp. 181–186. New York: Wiley.

Gomes, D. S., L. A. Andrade, C. J. N. Ribeiro, M. V. S. Peixoto, S. V. M. A. Lima, A. M. Duque, T. M. Cirilo, M. A. O. Góes, A. G. C. F. Lima, M. B. Santos, and et al. (2020). Risk clusters of covid-19 transmission in northeastern brazil: Prospective space–time modelling. *Epidemiology and Infection 148*, e188.

Gomez-Rubio, V. (2020). *Bayesian Inference with INLA* (1 ed.). New York: CRC Press.

Green, P. J. (1995). Reversible jump MCMC computation and Bayesian model determination. *Biometrika 82*, 711–732.

Green, P. J. and S. Richardson (2002). Hidden Markov models and disease mapping. *Journal of the American Statistical Association 97*, 1055–1070.

Gustafson, P. (2004). *Measurement Error and Misclassification in Statistics and Epidemiology*. London: Chapman & Hall.

Heagerty, P. and S. Lele (1998). A composite likelihood approach to binary spatial data. *Journal of the American Statistical Association 93*, 1099–1111.

Henderson, R., S. Shimakura, and D. Gorst (2002). Modeling spatial variation in leukaemia survival data. *Journal of the American Statistical Association 97*, 965–972.

Higdon, D. (2002). Space and space-time modeling using process convolutions. In C. Anderson and et al.(Ed.), *Quantitative Methods for Current Environmental Issues*. London: Springer.

Hoffman, M. D. and A. Gelman (2014). The no-u-turn sampler: Adaptively setting path lengths in Hamiltonian Monte Carlo. *Journal of Machine Learning Research 15*, 1351–1381.

Hossain, M. and A. Lawson (2008). Approximate methods in Bayesian point process spatial models. *Computational Statistics and Data Analysis 53*, 2831–2842.

Hossain, M. and A. B. Lawson (2005). Local likelihood disease clustering: Development and evaluation. *Environmental and Ecological Statistics 12*, 259–273.

Hossain, M. and A. B. Lawson (2006). Cluster detection diagnostics for small area health data: With reference to evaluation of local likelihood models. *Statistics in Medicine 25*, 771–786.

Hossain, M., A. B. Lawson, B. Cai, J. Choi, J. Liu, and R. S. Kirby (2012). Space-time stick-breaking processes for small area disease cluster estimation. *Environmental and Ecological Statistics*, 1–17.

Hossain, M. M. and A. B. Lawson (2010). Space-time Bayesian small area disease risk models: Development and evaluation with a focus on cluster detection. *Environmental and Ecological Statistics 17*, 73–95.

Hossain, M. M., A. B. Lawson, B. Cai, J. Choi, J. Liu, and R. S. Kirby (2014). Space-time areal mixture model: Relabeling algorithm and model selection issues. *Environmetrics 25*(2), 84–96.

Illian, J., H. Sorbye, and H. Rue (2012). A toolbox for fitting complex spatial point process models using integrated nested Laplace approximation (INLA). *Annals of Applied Statistics 6*(4), 1499–1530.

Inskip, H., V. Beral, P. Fraser, and P. Haskey (1983). Methods for age-adjustment of rates. *Statistics in Medicine 2*, 483–493.

Jasra, A., C. Holmes, and D. Stephens (2005). Markov Chain Monte Carlo methods and the label switching problem in Bayesian mixture modeling. *Statistical Science 20*, 50–67.

Jin, X., S. Banerjee, and B. Carlin (2007). Order-free coregionalized areal data models with application to multiple disease mapping. *Journal of the Royal Statistical Society B 69*, 817–838.

Jin, X., B. Carlin, and S. Banerjee (2005). Generalized hierarchical multivariate car models for areal data. *Biometrics 61*, 950–961.

Kawachi, I. and L. Berkman (Eds.) (2003). *Neighborhoods and Health*. New York: Oxford University Press.

Keeling, M. and P. Rohani (2007). *Modeling Infectious Diseases in Humans and Animals*. New York: Princeton University Press.

Keller, J., H. Chang, M. Strickland, and A. Szpiro (2017). Measurement error correction for predicted spatiotemporal air pollution exposures. *Epidemiology 28*, 338–345.

Kelsall, J. and J. Wakefield (2002). Modelling spatial variation in disease risk: A geostatistical approach. *Journal of the American Statistical Association 97*, 692–701.

Kibria, B., L. Sun, J. Zidek, and N. Le (2002). Bayesian spatial prediction of random space-time fields with application to mapping PM2.5 exposure. *Journal of the American Statistical Association 97*, 112–124.

Kim, J.-I., A. B. Lawson, S. McDermott, and C. Aelion (2010). Bayesian spatial modeling of disease risk in relation to multivariate environmental risk fields. *Statistics in Medicine 29*, 142–157.

Knorr-Held, L. (2000). Bayesian modelling of inseparable space-time variation in disease risk. *Statistics in Medicine 19*, 2555–2567.

Knorr-Held, L. and G. Rasser (2000). Bayesian detection of clusters and discontinuities in disease maps. *Biometrics 56*, 13–21.

Kuo, L. and B. Mallick (1998). Variable selection for regression models. *Sankhya 60*, 65–81.

Lambert, P., A. Sutton, P. Burton, K. Abrams, and P. Jones (2005). How vague is vague? A simulation study of the impact of the use of vague prior distributions in MCMC using WinBUGS. *Statistics in Medicine 24*, 2401–2428.

Lawson, A. and J. Choi (2016). Spatiotemporal disease mapping. In A. B. Lawson, S. Banerjee, R. Haining, and L. Ugart (Eds.), *Handbook of Spatial Epidemiology*, pp. 335–348. New York: CRC Press.

Lawson, A., A. Schritz, L. Villarroel, and G. A. Aguayo (2020). Multi-scale multivariate models for small area health survey data: A Chilean example. *International Journal of Environmental Research and Public Health 17*(5), 1682.

Lawson, A. B. (1992a). GLIM and normalising constant models in spatial and directional data analysis. *Computational Statistics and Data Analysis 13*, 331–348.

Lawson, A. B. (1992b). On fitting non-stationary Markov point processes on GLIM. In Y. Dodge and J. Whittacker (Eds.), *Computational Statistics I*. Physica Verlag.

Lawson, A. B. (2006). *Statistical Methods in Spatial Epidemiology* (2 ed.). New York: Wiley.

Lawson, A. B. (2012). Bayesian point event modeling in spatial and environmental epidemiology: A review. *Statistical Methods in Medical Research 21*, 509–529.

Lawson, A. B. (2018). *Bayesian Disease Mapping: Hierarchical Modeling in Spatial Epidemiology* (3 ed.). New York: CRC Press.

Lawson, A. B. (2020). Nimble for Bayesian disease mapping. *Spatial and Spatio-temporal Epidemiology 33*, 100323.

Lawson, A. B., A. Biggeri, D. Boehning, E. Lesaffre, J.-F. Viel, A. Clark, P. Schlattmann, and F. Divino (2000). Disease mapping models: An empirical evaluation. *Statistics in Medicine 19*, 2217–2242. special issue: Disease Mapping with emphasis on evaluation of methods.

Lawson, A. B., R. Carroll, and M. Castro (2014). Joint spatial Bayesian modeling for studies combining longitudinal and cross-sectional data. *Statistical Methods in Medical Research 23*, 611–624. DOI:10.1177/0962280214527383.

Lawson, A. B., R. Carroll, C. Faes, R. S. Kirby, M. Aregay, and K. Watjou (2017). Spatio-temporal multivariate mixture models for Bayesian model selection in disease mapping. *Environmetrics 28*.

Lawson, A. B., J. Choi, B. Cai, M. M. Hossain, R. Kirby, and J. Liu (2012). Bayesian 2-stage space-time mixture modeling with spatial misalignment of the exposure in small area health data. *Journal of Agricultural, Biological and Environmental Statistics 17*, 417–441.

Lawson, A. B. and N. Cressie (2000). Spatial statistical methods for environmental epidemiology. In C. R. Rao and P. K. Sen (Eds.), *Handbook of Statistics: Bio-Environmental and Public Health Statistics*, Volume 18, pp. 357–396. Elsevier.

Lawson, A. B. and J. Kim (2020). Space-time covid-19 Bayesian (SIR) modeling in South Carolina. submitted. (medRxiv 2020.11.03.20225227; doi: https://doi.org/10.1101/2020.11.03.20225227)

Lawson, A. B., G. Onicescu, and C. Ellerbe (2011). Foot and mouth disease revisited: Re-analysis using Bayesian spatial susceptible-infectious-removed models. *Spatial and Spatio-Temporal Epidemiology 2*, 185–194.

Lawson, A. B. and C. Rotejanaprasert (2018). Bayesian spatial modeling for the joint analysis of zoonosis between human and animal populations. *Spatial Statistics 28*, 8–20.

Lawson, A. B. and H. R. Song (2009). Semiparametric spacetime survival modeling of chronic wasting disease in deer. *Environmental and Ecological Statistics, 17*, 559–571. DOI 10.1007/s10651-009-0118-z.

Lawson, A. B. and H.-R. Song (2010). Bayesian hierarchical modeling of the dynamics of spatio-temporal influenza season outbreaks. *Spatial and Spatio-temporal Epidemiology 1*, 187–195.

Lawson, A. B., H.-R. Song, B. Cai, M. M. Hossain, and K. Huang (2010). Space-Time latent component modeling of geo-referenced health data. *Statistics in Medicine 29*, 2012–2017.

Lee, D. (2013). CARBayes: An R package for Bayesian spatial modeling with conditional autoregressive priors. *Journal of Statistical Software 55.*

Lee, D. and A. B. Lawson (2016). Quantifying the spatial inequality and temporal trends in maternal smoking rates in Glasgow. *Annals of Applied Statistics 10*(3), 1427–1446.

Lee, D. and C. Sarran (2015). Controlling for unmeasured confounding and spatial misalignment in long-term air pollution and health studies. *Environmetrics 26*, 477—-487.

Leroux, B., X. Lei, and N. Breslow (2000). Estimation of disease rates in small areas: A new mixed model for spatial dependence. In M. Halloran and D. Berry (Eds.), *Statistical Models in Epidemiology, the Environment and Clinical Trials*, pp. 135–178. New York: Springer-Verlag.

Lesaffre, E. and A. B. Lawson (2012). *Bayesian Biostatistics*. New York: Wiley.

Leyland, A. and H. Goldstein (Eds.) (2001). *Multilevel Modelling in Health Statistics*. London: Wiley.

Li, G., N. Best, A. L. Hansell, I. Ahmed, and S. Richardson (2012). BaySTDetect: Detecting unusual temporal patterns in small area data via Bayesian model choice. *Biostatistics 13*(4), 695–710.

Lindgren, F. and H. Rue (2015). Bayesian spatial modelling with R-INLA. *Journal of Statistical Software 63*, 1–25.

Lindgren, F., H. Rue, and J. Lindstrom (2011). An explicit link between Gaussian fields and Gaussian Markov random fields: The stochastic partial differential equation approach. *Journal of the Royal Statistical Society B 74*, 423–498.

Little, R. and D. Rubin (2019). *Statistical Analysis with Missing Data* (3 ed.). New York: Wiley.

Liu, J. S. (2001). *Monte Carlo Strategies in Scientific Computing*. New York: Springer.

Liu, X., M. Wall, and J. Hodges (2005). Generalised spatial structural equation models. *Biostatistics 6*, 539–557.

Lopes, H., E. Salazar, and D. Gamerman (2008). Spatial dynamic factor analysis. *Bayesian Analysis 3*, 759–792.

Louie, M. M. and E. D. Kolaczyk (2006). Multiscale detection of localised anomalous structure in aggregate disease incidence data. *Statistics in Medicine 25*, 787–810.

Lunn, D., C. Jackson, N. Best, A. Thomas, and D. Spiegelhalter (2012). *The BUGS Book*. New York: CRC Press.

MacNab, Y. C., A. Kmetic, P. Gustafson, and S. Sheps (2006). An innovative application of Bayesian disease mapping methods to patient safety research: A Canadian adverse medical event study. *Statistics in Medicine 25*(23), 3960–3980.

Marshall, E. and D. Spiegelhalter (2003). Approximate cross-validatory predictive checks in disease mapping models. *Statistics in Medicine 22*, 1649–1660.

Martinez-Beneito, M. A., P. Botella-Rocamora, and S. Banerjee (2017). Towards a multidimensional approach to Bayesian disease mapping. *Bayesian Anal. 12*(1), 239–259.

Mercer, L., J. Wakefield, C. Chen, and T. Lumley (2014). A comparison of spatial smoothing methods for small area estimation with sampling weights. *Spatial Statistics 8*, 69–85.

Miller, C., A. Lawson, D. Chung, M. Gebregziabher, E. Yeh, R. Drake, and E. Hill (2020). Automating a process convolution approach to account for spatial information in imaging mass spectrometry data. *Spatial Statistics 36*, 100422.

Moraga, P. and A. Lawson (2012). Gaussian component mixtures and car models in Bayesian disease mapping. *Computational Statistics and Data Analysis 56*, 1417–1433.

Morgan, O., M. Vreiheid, and H. Dolk (2004). Risk of low birth weight near EUROHAZCON hazardous waste landfill sites in England. *Archives of Environmental Health 59*, 149–151.

Morris, M., K. Wheeler-Martin, D. Simpson, S. J. Mooney, A. Gelman, and C. DiMaggio (2019). Bayesian hierarchical spatial models: Implementing the Besag York Mollie model in stan. *Spatial and Spatio-temporal Epidemiology 31*, 100301.

Morton, A. and B. Finkenstadt (2005). Discrete time modelling of disease incidence time series by using Markov Chain Monte Carlo methods. *Journal of the Royal Statistical Society C 54*, 575–594.

Napier, G., D. Lee, C. Robertson, and A. Lawson (2019). A Bayesian space-time model for clustering areal units based on their disease trends. *Biostatistics 20*, 681–697.

Neal, R. (2011). MCMC using Hamiltonian Dynamics. In S. Brooks, A. Gelman, G. Jones, and X. Meng (Eds.), *Handbook of Markov Chain Monte Carlo*, Chapter 5. New York: CRC Press.

Neal, R. M. (2003). Slice sampling. *Annals of Statistics 31*, 1–34.

Nieuwenhuisen, M. (2016). Environmental studies. In A. B. Lawson, S. Banerjee, R. Haining, and L. Ugart (Eds.), *Handbook of Spatial Epidemiology*, pp. 39–55. New York: CRC Press.

Nott, D. J. and T. Rydén (1999). Pairwise likelihood methods for inference in image models. *Biometrika 86*, 661–676.

O'Hara, R. B. and M. J. Sillanpää (2009). A review of Bayesian variable selection methods: What, how, and which. *Bayesian Analysis 4*, 85–118.

Okabe, A., B. Boots, and K. Sugihara (1992). *Spatial Tessellations*. New York: Wiley.

Onicescu, G. and A. B. Lawson (2016). Bayesian modeling and inference. In A. B. Lawson, S. Banerjee, R. Haining, and L. Ugart (Eds.), *Handbook of Spatial Epidemiology*, pp. 133–160. New York: CRC Press.

Onicescu, G., A. B. Lawson, S. McDermott, M. Aelion, and B. Cai (2014). Bayesian importance parameter modeling of misaligned predictors: Soil metal measures related to residential history and intellectual disability in children. *Environmental Science and Pollution Research 21*, 10775—10786.

Onicescu, G., A. B. Lawson, J. Zhang, M. Gebregziabher, K. Wallace, and J. M. Eberth (2017a). Bayesian accelerated failure time model for space-time dependency in a geographically augmented survival model. *Statistical Methods in Medical Research*. PMID:26220537.

Onicescu, G., A. B. Lawson, J. Zhang, M. Gebregziabher, K. Wallace, and J. M. Eberth (2017b). Spatially explicit survival modeling for small area cancer data. *Journal of Applied Statistics 0*, 1–18.

Osnes, K. and O. Aalen (1999). Spatial smoothing of cancer survival: A Bayesian approach. *Statistics in Medicine 18*, 2087–2099.

Petralias, A. and P. Dellaportas (2013). An MCMC model search algorithm for regression problems. *Journal of Statistical Computation and Simulation 83*(9), 1722–1740.

Raghunathan, T. E., D. Xie, N. Schenker, V. L. Parsons, W. W. Davis, K. W. Dodd, and E. J. Feuer (2007). Combining information from surveys to estimate county-level prevalence rates of cancer risk factors and screening. *Journal of the American Statistical Association 102*, 474–486.

Rao, J. K. N. (2003). *Small Area Estimation*. New York: Wiley.

Richardson, S., A. Thomson, N. Best, and P. Elliott (2004). Interpreting posterior relative risk estimates in disease mapping studies. *Environmental Health Perspectives 112*, 1016–1025.

Riebler, A., S. Sørbye, D. Simpson, and H. Rue (2016, 08). An intuitive Bayesian spatial model for disease mapping that accounts for scaling. *Statistical Methods in Medical Research 25*, 1145–1165.

Ripley, B. D. (1981). *Spatial Statistics*. New York: Wiley.

Ripley, B. D. (1987). *Stochastic Simulation*. New York: Wiley.

Robert, C. and G. Casella (2005). *Monte Carlo Statistical Methods* (2 ed.). New York: Springer.

Rockova, V. and E. George (2014). Negotiating multicollinearity with spike-and-slab priors. *Metron 72*, 217–229.

Rotejanaprasert, C., A. B. Lawson, H. Rossow, J. Sane, O. Huitu, H. Henttonen, and V. D. R. Vilas (2017). Towards integrated surveillance of zoonoses: Spatiotemporal joint modeling of rodent population data and human tularemia cases in Finland. *BMC Research Methodology*.

Rue, H. and L. Held (2005). *Gaussian Markov Random Fields: Theory and Applications*. New York: Chapman Hall/CRC.

Rue, H., S. Martino, and N. Chopin (2009). Approximate Bayesian inference for latent Gaussian models by using integrated nested Laplace approximations. *Journal of the Royal Statistical Society B 71*, 319–392.

Rue, H., A. Reibler, S. Sorbye, J. Illian, D. Simpson, and F. Lindgren (2016). Bayesian computing with INLA: A review. *arXiv*. https://arxiv.org/abs/1604.00860.

Sanden, A. and B. Jarvholm (1991). A study of possible predictors of mesothelioma in shipyard workers exposed to asbestos. *Journal of Occupational Medicine: 33*(7), 770—773.

Sattenspiel, L. and A. Lloyd (2009). *The Geographic Spread of Infectious Diseases: Models and Applications*. New York: Princeton University Press.

Schrodle, B. and L. Held (2011). Spatio-temporal disease mapping using INLA. *Environmetrics 22*, 725–734.

Schrodle, B., L. Held, A. Rieber, and J. Danuser (2011). Using integrated nested Laplace approximations for the evaluation of veterinary surveillance data from Switzerland: A case-study. *Applied Statistics 60*, 261–279.

Sibson, R. (1980). The Dirichlet tessellation as an aid in data analysis. *Scandinavian Journal of Statistics 7*, 14–20.

Simpson, D., F. Lindgren, and H. Rue (2012). In order to make spatial statistics computationally feasible, we need to forget about the covariance function. *Environmetrics 23*, 65–74.

Simpson, D. P., H. Rue, T. G. Martins, A. Riebler, and S. H. Sørbye (2017). Penalising model component complexity: A principled, practical approach to constructing priors. Statistical Science, 32, 1–28.

Smith, A. F. M. and A. E. Gelfand (1992). Bayesian statistics without tears: A sampling-resampling perspective. *American Statistician 46*, 84–88.

Smith, A. F. M. and G. Roberts (1993). Bayesian computation via the Gibbs Sampler and related Markov Chain Monte Carlo methods. *Journal of the Royal Statistical Society B 55*, 3–23.

Spiegelhalter, D., N. Best, B. Carlin, and A. van der Linde (2014). The deviance information criterion: 12 years on. *Journal of the Royal Statistical Society B 76*, 485–493.

Spiegelhalter, D. J., N. G. Best, B. P. Carlin, and A. van der Linde (2002). Bayesian deviance, the effective number of parameters and the comparison of arbitrarily complex models. *Journal of the Royal Statistical Society B 64*, 583–640.

Spiegelhalter, D. J., N. G. Best, W. R. Gilks, and H. Inskip (1996). Hepatitis B: A case study in MCMC methods. In W. R. Gilks, S. Richardson, and D. J. Spiegelhalter (Eds.), *Markov Chain Monte Carlo in Practice*. London: Chapman and Hall.

Stern, H. and Y. Jeon (2004). Applying structural equation models with incomplete data. In A. Gelman and X.-L. Meng (Eds.), *Applied Bayesian Modeling and Causal Inference from Incomplete-Data Perspectives*, Chapter 30, pp. 331–342. London: Wiley.

Stern, H. S. and N. A. C. Cressie (1999). Inference for extremes in disease mapping. In A. B. Lawson, A. Biggeri, D. Boehning, E. Lesaffre, J. F. Viel, and R. Bertollini (Eds.), *Disease Mapping and Risk Assessment for Public Health*, Chapter 5. New York: Wiley.

Stern, H. S. and N. A. C. Cressie (2000). Posterior predictive model checks for disease mapping models. *Statistics in Medicine 19*, 2377–2397.

Stevenson, M., R. Morris, A. B. Lawson, J. Wilesmith, J. M. Ryan, and R. Jackson (2005). Area level risks for BSE in British cattle before and after the July 1988 meat and bone meal feed ban. *Preventive Veterinary Medicine 69*, 129–144.

Tanner, M. A. (1996). *Tools for Statistical Inference* (3rd ed.). New York: Springer Verlag.

Taylor, B., T. Davies, B. Rowlingson, and P. Diggle (2015). Bayesian inference and data augmentation schemes for spatial, spatiotemporal and multivariate Log-Gaussian Cox processes in R. *Journal of Statistical Software 63*(7), 1–47.

Taylor, B. and P. Diggle (2014). INLA or MCMC? a tutorial and comparative evaluation for spatial prediction in Log-Gaussian Cox processes. *Journal of Statistical Computation and Simulation 84*(10), 2266–2284.

Thomas, A., N. Best, D. Lunn, R. Arnold, and D. Spiegelhalter (2004). GeoBUGS user manual v 1.2. www.mrc-bsu.cam.ac.uk/bugs.

Tibshirani, R. and T. Hastie (1987). Local likelihood estimation. *Journal of the American Statistical Association 82*, 559–568.

Tierney, L. and J. Kadane (1986). Accurate approximations for posterior moments and marginal densities. *Journal of the American Statistical Association 81*, 82–86.

Tzala, E. and N. Best (2008). Bayesian latent variable modelling of multivariate spatio-temporal variation in cancer mortality. *Statistical Methods in Medical Research 17*, 97–118.

Vandendijck, Y., C. Faes, R. Kirby, A. Lawson, and N. Hens (2016). Model-based inference for small area estimation with sampling weights. *Spatial Statistics 18, Part B*, 455 – 473.

VanderWeele, T. (2011). Causal mediation analysis with survival data. *Epidemiology 22*, 582–585.

VanderWeele, T. and E. Tchetgen (2017). Mediation analysis with time varying exposures and mediators. *Journal of the Royal Statistical Society: Series B 79*, 917–938.

Varin, C., G. Høst, and Ø. Skare (2005). Pairwise likelihood inference in spatial generalized linear mixed models. *Computational Statistics and Data Analysis 49*, 1173–1191.

Wackernagel, H. (2003). *Multivariate Geostatistics* (3rd ed.). New York: Springer.

Waller, L. and C. Gotway (2004). *Applied Spatial Statistics for Public Health Data*. New York: Wiley.

Waller, L. A., B. P. Carlin, H. Xia, and A. E. Gelfand (1997). Hierarchical spatio-temporal mapping of disease rates. *Journal of the American Statistical Association 92*, 607–617.

Wang, F. and M. Wall (2003). Generalized common spatial factor model. *Biostatistics 4*, 569–582.

Watjou, K., C. Faes, A. Lawson, R. S. Kirby, M. Aregay, R. Carroll, and Y. Vandendijck (2017). Spatial small area smoothing models for handling survey data with nonresponse. *Statistics in Medicine 36*(23), 3708–3745.

Wikle, C. K. (2002). Spatial modelling of count data: A case study in modelling breeding bird survey data on large spatial domains. In A. B. Lawson and D. G. T. Denison (Eds.), *Spatial Cluster Modelling*, Chapter 11. London: CRC press.

Williams, F., A. Lawson, and O. Lloyd (1992). Low sex ratios of births in areas at risk from air pollution from incinerators, as shown by geographical analysis and 3-dimensional mapping. *International Journal of Epidemiology 21*, 311–319.

Wolpert, R. L. and K. Ickstadt (1998). Poisson/gamma random field models for spatial statistics. *Biometrika 85*, 251–267.

Yu, B. and P. Mykland (1998). Looking at Markov samplers through cusum path plots: A simple diagnostic idea. *Statistics and Computing 8*, 275–286.

Zhou, H., A. B. Lawson, J. Hebert, E. Slate, and E. Hill (2008). Joint spatial survival modelling for the date of diagnosis and the vital outcome for prostate cancer. *Statistics in Medicine 27*, 3612–28.

Zhou, Y., L. Wang, L. Zhang, L. Shi, K. Yang, J. He, B. Zhao, W. Overton, S. Purkayastha, and P. Song (2020, 8). A spatiotemporal epidemiological prediction model to inform county-level covid-19 risk in the united states. *Harvard Data Science Review*. https://hdsr.mitpress.mit.edu/pub/qqg19a0r.

Zou, J., Z. Zhang, and H. Yan (2018). A hybrid hierarchical Bayesian model for spatiotemporal surveillance data. *Statistics in Medicine 37*(28), 4216–4233.

Index

Printed in the United States
By Bookmasters